3rd International Congress of Mechanical Engineering and Agricultural Science (CIIMCA 2017)

IOP Conference Series: Materials Science and Engineering Volume 437

Bucaramanga, Columbia
2 – 6 October 2017

ISBN: 978-1-5108-7655-2
ISSN: 1757-8981

Printed from e-media with permission by:

Curran Associates, Inc.
57 Morehouse Lane
Red Hook, NY 12571

Some format issues inherent in the e-media version may also appear in this print version.

Printed by Curran Associates, Inc. (2019)

For permission requests, please contact the Institute of Physics
at the address below.

Institute of Physics
Dirac House, Temple Back
Bristol BS1 6BE UK

Phone: 44 1 17 929 7481
Fax: 44 1 17 920 0979

techtracking@iop.org

Additional copies of this publication are available from:

Curran Associates, Inc.
57 Morehouse Lane
Red Hook, NY 12571 USA
Phone: 845-758-0400
Fax: 845-758-2633
Email: curran@proceedings.com
Web: www.proceedings.com

TABLE OF CONTENTS

IOP Conf. Series: Materials Science and Engineering **437** (2018) 011001 doi:10.1088/1757-899X/437/1/011001

3rd International Congress of Mechanical Engineering and Agricultural Science (CIIMCA 2017)

Preface. The Mechanical Engineering Department of the Universidad Pontificia Bolivariana Bucaramanga (Colombia) organized the 3rd International Congress of Mechanical Engineering and Agricultural Science (CIIMCA 2017), from October 2 to 6, 2017. This congress made part of the International Week of Research, Development and Innovation (SIIDI 2017), which took place in the University Campus and included the 2nd international Conference Science and Technology of Drying (CiCTS 2017) and the 3rd International meeting of the RIIVI net.

Participants were invited to present original research works dealing with 5 thematic areas: 1) Science and Technology of Drying, 2) Engineering fundamentals, modelling, testing and simulation, 3) Renewable energy and thermal science, 4) Material science and engineering, 5) Automation and Agricultural Science. During the event oral and poster sessions were arranged according to these thematic areas.

The authors of all communications were also invited to submit an extended version of their research works to be published in a volume of the IOP Conference Series Material Science and Engineering or in the UIS Engineering Journal. Those contributions that successfully completed the editorial evaluation and a peer reviewed process, form part of this publication. The editorial team and Congress Organizers would like to express the members of the Scientific Committee and Peer Evaluators our sincere gratitude for their invaluable help and commitment.

More information at http://ciimca.upbbga.edu.co.

The editorial team.

Scientific committee

Dr. Ing. Gilles Gauthier	Colombia - Universidad Industrial de Santander - Chemical Engineering Department
Dr. Ing. Douglas Bressan Riffel	Brazil - Universidade Federal de Sergipe - Mechanical Engineering Department
Dr. Prof. Ing.Ching Lik Hii	Malaysia - University of Nottingham - Faculty of Engineering
Dr. Ing. Kento Taro Magara	Colombia - Universidad Pontificia Bolivariana- Faculty of Enviromental Engineering
Dr. Ing. Omar Pinzón Ardila	Colombia --Universidad Pontificia Bolivariana -Faculty of Electronic Engineering
Dr. Ing. Carlos Julio Cortes	Colombia - Universidad Nacional de Colombia -- Mechanical Engineering Department
Prof. Dr. Ing. Da-Wen Sun	Ireland - University College Dublin -Food Refrigeration and Computerised Food Technology Centre
Prof. Dr. Aron Mujumdar	Canada - McGill University
Pr. Dr. Oliver Hensel	Germany - University of Kassel - Ecological and Agricultural Sciences Department
Dr. Ing. Ma. Paola Maradei	Colombia - Universidad Industrial de Santander - Chemical Engineering Department
Pr. Dr. Ing. Adrian Mitroi	Romania - University of Agronomic Sciences and Veterinary Medicine Bucharest
Dr. Ing. Anjum Munir	Pakistan - University of Agriculture Faisalabad - Faculty of Agricultural Engineering & Technology
Dr. Ing. Sonia Rincon Prat	Colombia - Universidad Nacional de Colombia - Faculty of Mechanical and Mechatronic engineering
Dr. Ing. Ma. Ximena Quintanilla	Colombia -Universidad de la Sabana -Agroindustrial production engineering
Prof. Dr. Ing. Raghu Echempati	United States of America -Kettering University - Mechanical Engineering Department
Dr. Ing. Miguel Altuve Paredes	Venezuela - Universidad Pontificia Bolivariana Faculty of Electronic Engineering
Ms. Ing. Urbano Eliecer Prada	Colombia- Universidad Pontificia Bolivariana Faculty of Computer Science Engineering
Dr. Ing. Rogerio Gomes de Oliveira	Brazil - Federal University of Santa Catarina Energy and Sustainability Department
Dr. Ms. Diana Sinuco	Colombia Universidad Nacional de Colombia - Chemical Department
Dr. Roberto Alonso Gonzalez Lezcano	España – Universidad CEU San Pablo.-.Architecture and Design Department
Dr. Ing. Gail Gutierrez	Colombia – Universidad de la Guajira - Mechanical Engineering Department

IOP Conf. Series: Materials Science and Engineering **437** (2018) 011001 doi:10.1088/1757-899X/437/1/011001

Disclaimer

Facts and opinions in articles published on this CIIMCA 2017 Volume are solely the personal statements of respective authors. Authors are responsible for all contents in their article(s) including accuracy of the facts, statements, citing resources, and so on. The editorial team disclaim any liability of violations of other parties' rights, or any damage incurred as a consequence to use or apply any of the contents of this publication.

IOP Conf. Series: Materials Science and Engineering **437** (2018) 011002 doi:10.1088/1757-899X/437/1/011002

Peer review statement

All papers published in this volume of *IOP Conference Series: Materials Science and Engineering* have been peer reviewed through processes administered by the proceedings Editors. Reviews were conducted by expert referees to the professional and scientific standards expected of a proceedings journal published by IOP Publishing.

IOP Conf. Series: Materials Science and Engineering **437** (2018) 012001 doi:10.1088/1757-899X/437/1/012001

Adaption of the Cumulated Thermal Load for the Investigation of the Quality of Dried Agricultural Products

K Jödicke[1], S Arendt[1,2], W Hofacker[1] and W Speckle[2]

[1] Institute of Applied Thermo- and Fluiddynamics, University of Applied Sciences Konstanz Alfred-Wachtel-Straße 8, 78462 Konstanz, Germany
[2] Environmental Analysis Laboratory, University of Applied Sciences Ravensburg-Weingarten Doggenriedstraße, 88250 Weingarten, Germany

katrin.joedicke@htwg-konstanz.de

Abstract. When designing drying processes for sensitive biological foodstuffs like fruit or vegetables, energy and time efficiency as well as product quality are gaining more and more importance. These all are greatly influenced by the different drying parameters (e.g. air temperature, air velocity and dew point temperature) in the process. In sterilization of food products the cooking value is widely used as a cross-link between these parameters. In a similar way, the so-called cumulated thermal load (CTL) was introduced for drying processes. This was possible because most quality changes mainly depend on drying air temperature and drying time. In a first approach, the CTL was therefore defined as the time integral of the surface temperature of agricultural products. When conducting experiments with mangoes and pineapples, however, it was found that the CTL as it was used had to be adjusted to a more practical form. So the definition of the CTL was improved and the behaviour of the adjusted CTL (CTL_{ad}) was investigated in the drying of pineapples and mangoes. On the basis of these experiments and the work that had been done on the cooking value, it was found, that more optimization on the CTL_{ad} had to be done to be able to compare a great variety of different products as well as different quality parameters.

1. Introduction

Energy consumption, time efficiency and product quality are the most important aspects in the design of drying processes for sensitive biological foodstuffs like pineapples or mangoes. The outcome of different drying processes with perhaps even varying drying parameters (drying temperatures, air velocities, drying time and humidity), however, is very hard to compare. Due to this, the cumulated thermal load has been developed to improve the comparability of processes with different drying parameters [1]. In a similar way as the cooking value in sterilization [2], the so-called CTL combines the drying time with the product's surface temperature during the drying process, and therefore makes it possible to easily compare drying processes with different temperatures and drying times.

In detail, the CTL is a more basic and also more general value than the cooking value, which has to be defined differently for different products and different processes. It is defined as the time integral of the surface temperature of the agricultural products dried over the drying time. By achieving a low CTL, a minimum of energy consumption can be achieved as well, since both time and temperatures have the biggest impact on a drying process' energy consumption. Since a lot of quality parameters (e. g. colour changes, mechanical properties like texture and shrinkage during drying, contents of temperature sensitive acids as well as sugar contents) vary with both temperature and time, the CTL has been proven to be useful to investigate the impact of time and temperature on those changes and to find optimal drying conditions [3], [4], [5], [6]. For an optimal use of the CTL, however, some improvements had to be made. This work describes the successful first step of adjusting the CTL, but then focusses on further necessary changes in the adjusted CTL_{ad}.

2. Materials and Methods

2.1. Cooking value and CTL

The cooking value c is used in sterilization and pasteurization to describe the amount of heat a processed product receives over the whole processing time t. It takes into account the degradation of nutrients, changes in texture of the products as well as microbiological and enzymatic reactions. Its mathematical definition is [2]:

$$C_{T_{ref}}^{z_c} = \int_0^t 10^{\frac{T-T_{ref}}{z_c}}\,dt \tag{1}$$

with t = processing time
T = temperature
T_{ref} = reference temperature
z_c = thermal destruction rate (for microbial inactivation)

The cooking value is dependent on 2 reference values: z_c which is dependent on the heat-vulnerable components (e. g. bacteria, vitamins, colour, texture) and T_{ref} which is generally set as 100 °C or 121.1 °C (250 °F). Detailed numbers for both reference values can be found in literature [2].

The CTL has the same basic idea as the cooking value -measuring the thermal load of a process on the processed good- but puts it in a very basic way. It is defined as:

$$CTL = \int_0^t T_S\,dt \tag{2}$$

with t = processing time
T_S = surface temperature

Despite this simple approach, the use of the CTL still was possible, since it was accepted that non-linear changes in quality criteria during the process would not be taken into account. This was also possible because very good results could be obtained in experimental work with apples [1]. It thus has proven to be a simple yet highly informative parameter. Figure 1 shows an example of a CTL for a drying process with an air temperature of 60 °C and a drying time of about 500 min.

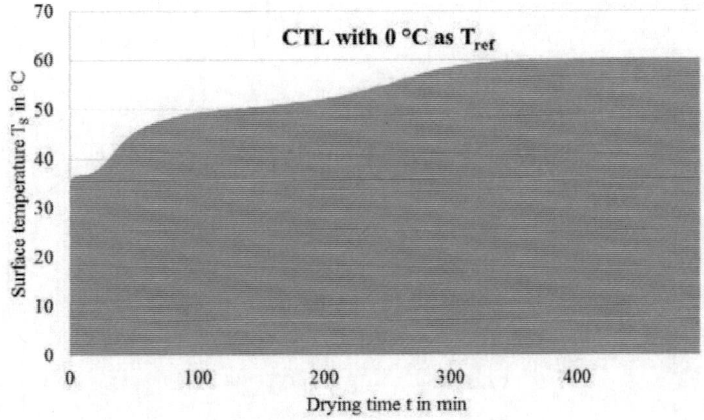

Figure 1. Original CTL

2.2. Adjusting of CTL

Despite the good results in previous works, an optimization of the CTL seemed useful and therefore a new adjusted CTL (CTL_{ad}) was developed. The adjusting of the CTL was done in a theoretical approach and partially confirmed in experiments.

In a first step, the units of the CTL_{ad} were changed from °C and min to the standard metric units K and s. Then the temperature data was flattened using a Gaussian normal distribution. Like this, the data was prepared for the calculation of the CTL. The main focus of the work, however, was to find a more reasonable reference temperature of the CTL than the 0 °C that was used previously. The influence of various reference temperatures (like the ambient air temperature, the dew point temperature, the air temperature for drying, the initial temperature of the drying experiment, as well as combinations of those temperatures) on the CTL was analysed. A new reference temperature was then chosen and results of the quality analysis of the drying experiments were compared using this new CTL_{ad}. Since these steps still did not produce a CTL_{ad} that was useable for all products or quality criteria, further investigations were and still are done on how to find a good balance between drying time t and surface temperature T_S.

2.3. Drying Process

A total of 32 randomized drying runs for pineapples and 17 for mangoes were planned and carried out, with 17 runs being the minimum amount of runs required for an adequately high reliability of the design.

All drying experiments were conducted at a through-flow single-layer dryer. The dryer's setup is shown in Figure 2. A fan (1) blows air with a settable air velocity v_{air} through the dryer. The air then enters a water bath with controlled dew point temperatures T_{DP}. They vary between 5 °C and 40 °C (2). After leaving the water bath, the air has a humidity of 100 %. The air is then heated up to air temperatures T_{air} between 40 °C and 94 °C (3). This leads to relative humidities RH of 2.5 % to 58 %. The air is then vertically blown through a grid on which a single layer of product is placed for drying (4). At this point the velocity of the air lies between 0.4 m/s and 1.4 m/s. The air and the surface temperature of the product (T_S) and the weight of the samples are continuously measured in the drying chamber (5) and photos of the products are taken with a CCD camera for colour and texture analysis (shrinkage).

IOP Conf. Series: Materials Science and Engineering **437** (2018) 012001 doi:10.1088/1757-899X/437/1/012001

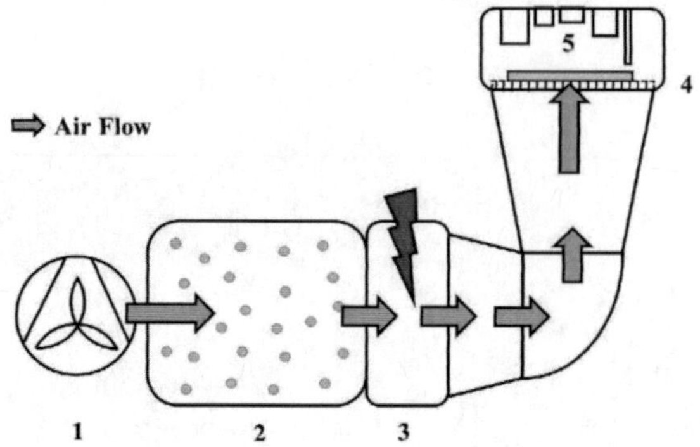

Figure 2. Single Layer Dryer

2.4. Samples and Sample Preparation

For all experiments, Rothschild / Cayenne pineapples and Kent mangoes were used. To guarantee a stable quality of the fresh products, all fruit were purchased at a local market. Thus, changes from the use of different products or types of fruit could be eliminated as far as possible. This was very important for being able to only measure the impact of the drying parameters on quality rather than the changes in the products themselves.

For the pineapple experiments, only a single pineapple had to be used per run. All pineapples were peeled. To obtain equally thick slices of $s = 8$ mm, a *Graeff C20EU* slicing machine was used for cutting the pineapple samples. To obtain equal samples for drying and chemical analysis (Figure 3), the slices were cut by using round punches ($d_{out} = 100$ mm and $d_{in} = 35$ mm) and a knife.

Figure 3. Shape of pineapple slices for fresh and dry chemical analysis(left) and texture and colour (right)

The quarters (left in Figure 3) were used for drying and for chemical analysis. The two eighths (right in Figure 3) were used for colour analysis during the drying process and for texture analysis after drying. The slices were thoroughly mixed before chemical analysis and drying. This step is crucial, since pine-apples as well as mangoes, have "sunny" and "shady" sides, between which the contents of vitamins and sugars vary.

The dryer was preheated to the chosen temperature, humidity and air velocity. Approximately 2/3 of the slices (equalling 260 g / run) were placed on the grid in the dryer. The grid with the pineapples slices is shown in Figure 4. A black background and a white standard are necessary for colour analysis. The surface temperature T_S is continuously measured by a pyrometer, the measuring position is indicated by the red laser dot on the right.

Figure 4. Pineapple slices, white standard and laser dot of pyrometer for T_S measurement on drying grid

About 1/3 of the slices were used for fresh chemical analysis. The samples, which were frozen with liquid nitrogen, were grinded in a *Moulinex* chopper. This way, an evenly distributed fine powder could be obtained for each experiment. The powders were kept frozen until the chemical analysis took place. A small amount of the fresh products was not frozen but directly used for the refractive determination of the °Brix value of the fresh samples with a *Kruess DR101-60* refractometer.

The samples for drying were dried to a mass constancy of +/- 0.1 g over 30 min which leads to final moisture contents after drying of less than 20 %. They were then removed from the grid. The 2 eighths for texture analysis were taken and the rest of the dried samples were frozen with liquid nitrogen and grinded for chemical analysis. The powder obtained was again kept frozen until it was chemically analysed.

For the mango experiments, two mangoes had to be used per run. The preparation procedure was very similar to that of the pineapples. The main difference was the size and shape of the samples: they had a thickness of s = 4 mm and a round shape with d_{out} = 20 mm. The round shape was chosen, so less product of each mango would go to waste, and in a round of preliminary experiments, thinner slices had been proven to be more suitable for mangoes regarding drying kinetics. Because thinner and smaller slices were used, the total amount of product placed on the grid was also smaller (in total 150 g / run). The drying process itself and the preparation for the chemical analysis were exactly the same as described for the pineapples. Figure 5 shows the arrangement of the mango slices on the grid.

Figure 5. Mango slices, white standard and laser dot of pyrometer for T_S measurement on drying grid

2.5. Analysis of Colour and Mechanical Properties

During the drying process, the following variables were measured at a rate of one measurement per minute: total mass of samples, air temperature T_{air}, dew point temperature T_{DP} and surface temperature T_S. A photo of the samples with the white reference and the black background for colour analysis was taken in adequate time intervals (5 – 10 min), depending on the overall drying time. With the "ColorAnalyzer" software of the Institute of Applied Thermo- and Fluiddynamics IATF the L*, a* and b* values for one sample per run were calculated in each image taken. To analyse and interpret the measured data, the total colour difference ΔE and the ΔL*, Δa* und Δb* values of the samples between start and end of the drying were determined. The 2D-shrinkage was determined by counting and comparing the number of pixels (= size of the sample) analysed in each picture taken.

2.6. Chemical Analysis

Chemically, the retentions of ascorbic acids (vitamin C) and polyphenols were investigated. To be able to compare the results, they had to be set in relation to the dry matter content. In a first step, the sample powder was homogenized and then extracted. For measuring the ascorbic acid content, both, fresh and dried products, were prepared according to the European Standard EN 14130. The ascorbic acid extracts of the fresh and dried samples were analysed with a *1260 Infinity, AGILENT TECHNOLOGIES HPLC*. The sample components were separated using a *LiChrosher: 5 µ C18 100 A* reversed phase column with a dimension of 4.0 x 250 mm. The analysing of the ascorbic acid was done by isocratic elution. An eluent consisting of ultrapure water with KH_2PO_4 and methanol with N-cetyl-N,N,N-trimethyl-ammonium bromide was used (European Standard EN 14130). The detection of the ascorbic acid was carried out at a wavelength of 265 nm using a diode-array-detector [7], [8].

For the polyphenol analysis, the powders were weighed to 1.000 ± 0.005 g and were then extracted within 120 minutes at room temperature in 20 ml of methanol and 1 % acetic acid at a ratio of 80 : 20 (v / v). The samples were filtered and diluted with a solvent to 25 %. 500 µl of each sample extract were mixed with 500 µl of Folin-Ciocalteu reagent. After a reaction time of 5 minutes, 4000 µl of a 5.3 % sodium carbonate solution were added. After another 90 minutes the spectrometric absorption was measured at 756 nm against the blank.

3. Results and Discussion

3.1. Flattening of temperature data

To prepare the data for the calculation of the CTL_{ad}, the measured surface temperature values were flattened. This was done to reduce the uncertainty of the measured values. Previously, a simple arithmetic average out of 5 values was taken. In a new approach, the use of a Gaussian normal distribution was implemented. The 7 weights that were given to the temperature values (g_n to g_{n-6}) were calculated out of the standard normal distribution with the function shown in formula 3. The calculation of the flattened T_S is shown in formula 4. Table 1 gives an overview of all calculated values.

$$g = f(x) = \frac{1}{\sqrt{2\pi\sigma^2}} e^{-\frac{(x-\mu)^2}{2\sigma^2}} \tag{3}$$

with g = weight given to temperature value
µ = mean or expectation of the distribution (0 in a Gaussian standard normal distribution)
σ^2 = variance (1 in a Gaussian standard normal distribution)

$$T_{S\,flattened} == \frac{(T_n * g_n) + (T_{n-1} * g_{n-1}) + (T_{n-2} * g_{n-2}) + \cdots + (T_{n-6} * g_{n-6})}{g_n + g_{n-1} + g_{n-2} + \cdots + g_{n-6}} \tag{4}$$

with T_n = surface temperature T_S at point of time n
$T_{n-1\,...\,-6}$ = surface temperature T_S at point of time n-1 … -6
g_n = weight given to surface temperature T_S at point of time n
$g_{n-1\,...\,-6}$ = weight given to surface temperature T_S at point of time n - 1 … -6

Table 1. Gaussian flattening for б2=1, μ=0

Surface temperature T_S	x	Value of Gaussian normal dis-tribution
T sat t_{n-6}	-3	g_{n-6}=0,0044
T_S at t_{n-5}	-2.5	g_{n-5}=0,0175
T_S at t_{n-4}	-2	g_{n-4}=0,0539
T_S at t_{n-3}	-1,5	g_{n-3}=0,1295
T_S at t_{n-2}	-1	g_{n-2}=0,2419
T_S at t_{n-1}	-0,5	g_{n-1}=0,3520
T_S at t_n	0	g_n=0,3989

With the use of this weighting , it was possible to give a higher importance to the value at the point t=n and gradually lower values to points of time that were farther past (t=n-1 … t=n-6). Values in the future were not considered at all, since they don't have any impact on the process at point t=n. This flattened temperature data was then used for the calculation of the adjusted cumulated thermal load.

3.2. *Basic Adjusting of CTL_{ad}*

In a first step the new CTL_{ad} was then defined as:

$$CTL_{ad} = \sum_{n-1}^{N} \frac{(T_{n-1} - T_{ref}) + (T_n - T_{ref})}{2} * \Delta t \quad (5)$$

with
T_{n-1} = surface temperature T_S at point of time n-1
T_n = surface temperature T_S at point of time n
T_{ref} = reference temperature
t = time

After analysing different temperatures as T_{ref}, the initial surface temperature of the products $T_{S\ Start}$, recorded at the beginning of the experiment (t = 0 s), was chosen as the reference temperature. The effect of this on the CTL can be seen in Figure 6: by eliminating the area with no changes underneath $T_{S\ Start}$ (in this example 35 °C), the influence of changes in the CTL_{ad} on the overall CTL_{ad} gets bigger.

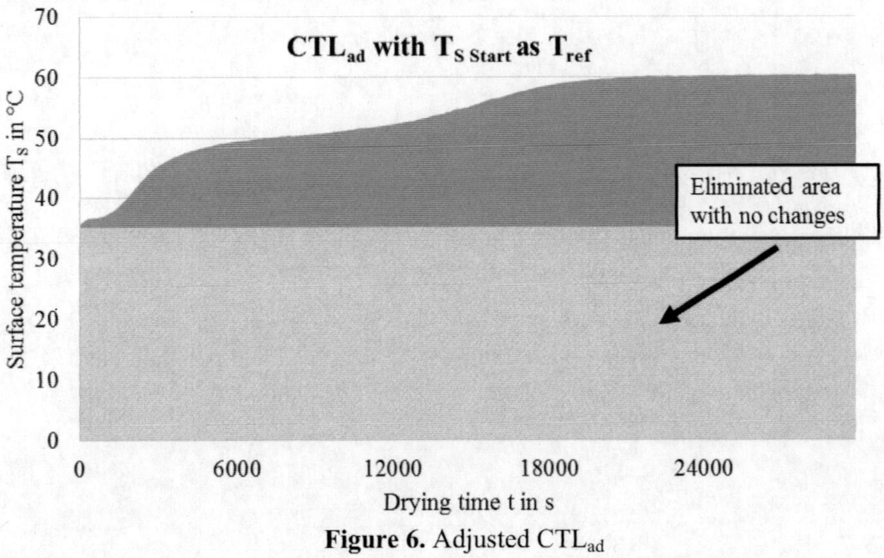

Figure 6. Adjusted CTL_{ad}

3.3. Verification through Experimental Results

With the help ofthe CTL_{ad}, the drying experiments on mangoes and pineapples were examined regarding changes in colour (ΔE), vitamin C retention (ascorbic acid) and polyphenol retention. All these quality parameters strongly depend on the drying parameters used. The desired results for the quality parameters were defined as shown in Table2.

Table 2. Optimum values for quality parameters used

Quality parameter	Unit	Optimum value	Explanation
Retention vitamin C (ascorbic acid)	%	As close to 100% as possible	As little change in taste and nutritious value as possible
Retention polyphenols	%	As close to 100%as possible	As little change in taste and nutritious valueas possible
Total colour difference ΔE	-	minimum	As little change in colour as possible
CTL_{ad}	$K*s*10^5$	minimum	Minimum of energy consumption and minimum of thermal impact on the product

The following figures show the results of the drying experiments. Figure 7 shows the retention of vitamin C (ascorbic acid) in pineapples over both the original CTL (left) and the adjusted CTL_{ad} (right). It shows the quality criteria of the same experiments over the differently calculated CTLs. When comparing the two, one can clearly see that the original CTL strongly depends on the temperature rather than on the "mix" of temperature and time. This is shown by the red dotted lines. The use of the CTL_{ad} clears away this strong effect of only temperature and makes it possible to find areas of optimal results for the combination of both parameters. This is shown as the red circle. The optimal areas described in Table 2 could be found by the use of this CTL_{ad} as shown and described in the following Figure 7 and Figure 8.

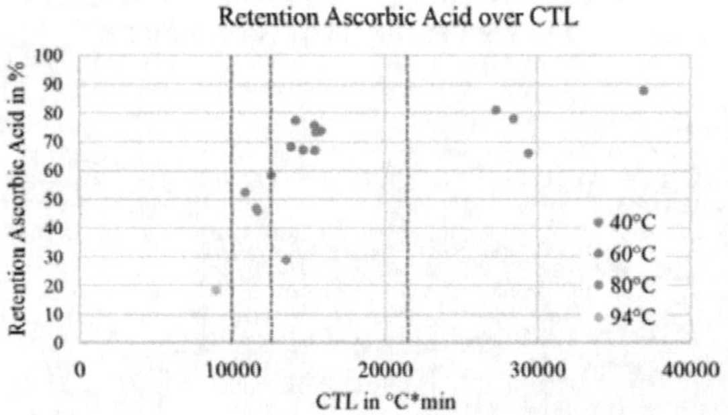

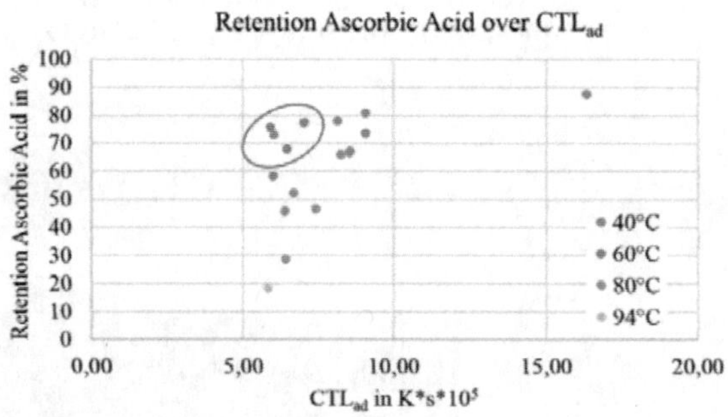

Figure 7. Retention of vitamin C (ascorbic acid) in pineapples over CTL (left) and CTL_{ad} (right) with T_{air} as additional parameter

IOP Conf. Series: Materials Science and Engineering **437** (2018) 012001 doi:10.1088/1757-899X/437/1/012001

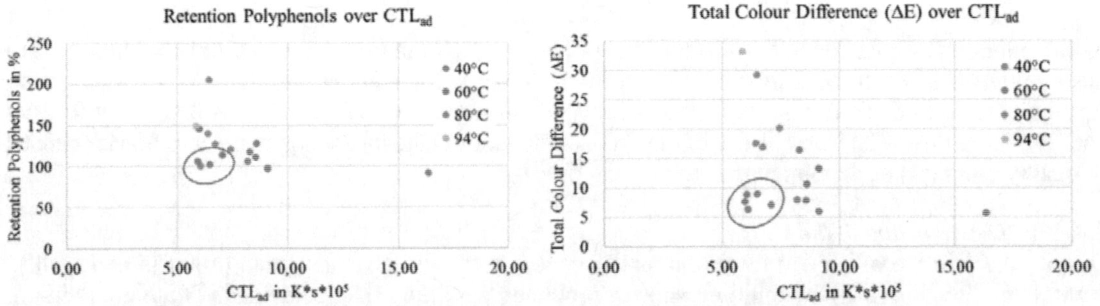

Figure 8. Retention of polyphenols (left) and ΔE (right) in pineapples over CTL_{ad} with T_{air} as additional parameter

With the help of these graphs and statistical software, it was already possible to find optimum drying parameters regarding energy consumption (drying time and temperatures) as well as quality. In the case of pineapple drying, a retention of vitamin C (ascorbic acid) and polyphenols around 100 %, with at the same time a low ΔE and low CTL_{ad} could be achieved at $T_{air} = 60$ °C, $v_{air} = 0.9$ m/s and RH = 12 % ($T_{DP} = 20$ °C). Up to now, however, only the differences between the initial values and the values after the total drying time were analysed and no closer investigation of the behaviour during the overall drying was done.

The results for the mango drying experiments are shown in Figure 9 and Figure 10. A similar tendency as for the pineapples can be seen. The effect of temperature on the CTL for mangoes is not as high as it is on pineapples, but as it can be seen in Figure 9 on the left, no real "mix" of T and t was possible when using the original CTL. This combined effect could be obtained by using the CTL_{ad}. With the help of the CTL_{ad} it was again possible to find the best results for the retention of vitamin C (ascorbic acid), retention of polyphenol and the total colour difference ΔE. Those optimal areas are again shown with the help of the red circles.

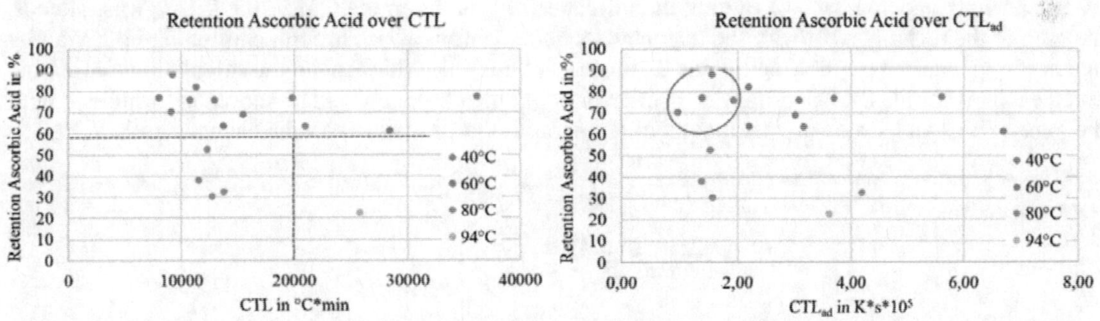

Figure 9. Retention of vitamin C (ascorbic acid) in mangoes over CTL (left) and CTL_{ad} (right) with T_{air} as additional parameter

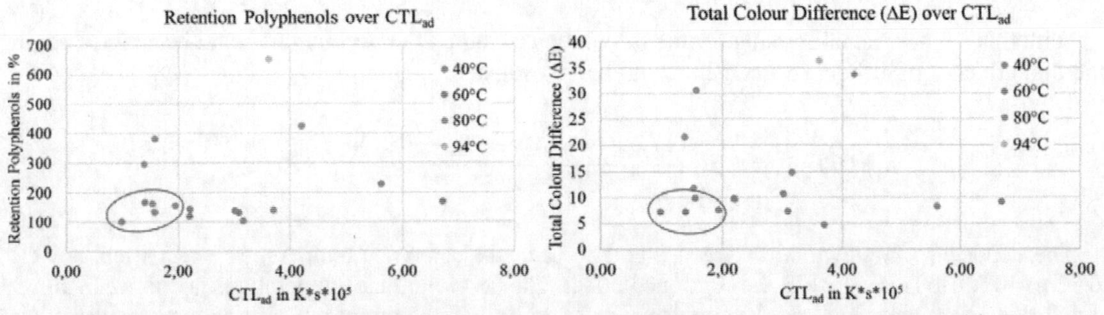

Figure 10. Retention of polyphenols (left) and ΔE (right) in mangoes over CTL_{ad} with T_{air} as additional parameter

In the same way as it was done for pineapples, optimal results for mangoes could also be found by using these graphs and statistical software. The optimum drying parameters regarding energy consumption (drying time and temperatures), as well as quality for mangoes, would be at T_{air} =40°C, v_{air} = 0.9 m/s and RH = 3 % (T_{DP} = 20 °C). The use of the CTL_{ad} again didn't take into account any non-linear changes of quality criteria at each point during the drying process.

3.4. Further Adjusting of the CTL_{ad}

Despite the positive effect of this first improvement and the fact that for some products and quality parameters its positive effect could be shown, problems with the CTL_{ad} remained. The biggest issues were first the fact that it was still not sure if the new reference temperature was the best one. Especially the fact that for each run a new reference temperature was used seemed to be not optimal. Further, the non-linear behavior of some quality criteria was not taken into account by the CTL_{ad} since only the values before and after drying were compared. The third problem was the fact that for long drying times, the effect of the drying temperature on the quality of the product is low and for short drying times, the effect of the drying temperature on the quality of the product is very high. The same could be applied the other way round: the higher the drying temperature, the lower the effect of time and the lower the drying temperature the higher the effect of time on the product quality. Therefore further improvements on the CTL_{ad} were made. It was decided to start those improvements with a variation of the surface temperature and the correlation between t and T_S and not yet look at the non-linear behavior of quality criteria over the drying time. Formula 6 again shows the basic calculation of the CTL_{ad}.

$$CTL_{ad} = \sum_{n-1}^{N} \frac{(T_{n-1} - T_{ref}) + (T_n - T_{ref})}{2} * \Delta t = \frac{1}{2}T_1 + \frac{1}{2}T_N + \sum_{n=2}^{N-1} T_n \tag{6}$$

For a very basic approach of finding the influence of T and t on the CTL_{ad}, the CTL_{ad} was taken as the area of the rectangle T*t with the exponents x and y (formula 7). Through combining both x and y in a new exponent x* the use of only one exponent was possible which made the calculation and testing much easier (formula 8). The propagation of error could then be calculated as shown in formula 9, with the exponents x and y showing the impact of temperature (x) or time (y) on the changes in the CTL_{ad}.

$$CTL_{ad} = (T - T_{ref})^{x} * t^{y} \tag{7}$$

$$CTL_{ad}^{*} = (T - T_{ref})^{x^*} * t \tag{8}$$

$$\frac{\Delta CTL_{ad}}{CTL_{ad}} = x\frac{\Delta(T - T_{ref})}{(T - T_{ref})} + y\frac{\Delta t}{t} = x^*\frac{\Delta(T - T_{ref})}{(T - T_{ref})} + \frac{\Delta t}{t} \tag{9}$$

With this x* the original formula for the calculation of the CTL_{ad} (formula 6) could be adapted again and the influence of different x* and T_{ref} could be investigated:

$$CTL_{ad}^{*} = \sum_{n-1}^{N} \left(\frac{(T_{n-1} - T_{ref}) + (T_n - T_{ref})}{2}\right)^{x^*} * \Delta t \tag{10}$$

The exponent x^* was varied between 0 and 5. Later the reference temperature was varied as well (0° C to 40 °C) to have a look at the combined effect. The best reference temperatures lay between 20 °C and 26 °C. Those gave good results since they were as close to the minimum starting temperature of all experiments as possible. Like this, the rectangular area that could be eliminated (compare: Figure 6) is as big as possible. Temperatures higher than the minimum starting temperature of all experiments would give no results for experiments that start with a lower temperature than the reference temperature. Those temperatures can therefore not be used. The optimal reference temperature in this work's pineapple drying experiments lay at 22 °C. For mangoes it lay at 26 °C. When investigating the exponents x^*, optimal regions of x^* regarding a mixed influence of T and t could be found in a first approach for the experiments conducted: for pineapples they lay between 0.8 – 1.1 and 1.6 – 3.1 and for mangoes between 0.4 – 0.8 and 1.3 -1.6. The detailed meaning of these x^* is currently being investigated.

4. Conclusions

The use of the adjusted CTL_{ad} for the comparison of different drying experiments is highly recommended, since a rise in comparability between different experiments could be shown regarding energy consumption and quality. The first very basic theoretical approach was proven to be useful by a round of experiments with mangoes and pineapples. Especially for the retention of vitamin C (ascorbic acid), polyphenols and ΔE, good results for mangoes as well as for pineapples could be obtained.

But since it is also clear, that, in a similar way as the cooking value that is used in sterilization processes, the CTL_{ad} varies differently with each quality parameters analysed and each product used, further adaptions of the CTL_{ad} were investigated. Possibilities for this further improvement of the CTL_{ad} lay in the fact that for short drying times, the effect of air temperature on quality changes is higher than the one of time and vice versa. To investigate this link, the use of different mathematical operations for the calculation of the CTL_{ad} was tested. During this testing the use of other reference temperatures in combination with the mathematical operations was also investigated. The best results in the approach, so far, came from a reference temperature of 22 °C and x* between 0.8 – 1.1 and 1.6 – 3.1 for pineapples and from a reference temperature of 26 °C and x* between 0.4 – 0.8 and 1.3 -1.6 for mangoes. The meaning of the results is still being investigated and has to be confirmed by more experiments with mangoes and pineapples, as well as with other products. Further, more quality parameters will be investigated, such as retentions in malic acids, citric acids, sugar contents, or changes in UV- or IR- spectra. Especially when non-linear behaviour of quality parameters exists, the CTL_{ad} needs further improvement. For a deeper insight into those non-linear quality criteria, a connection with the sterilization and pasteurisation process and the cooking value will be analysed.

5. Notation

a*	Red-green axis in the L*a*b* colour space
b*	Yellow-blue axis in the L*a*b* colour space
c	Cooking value
CTL	Cumulated thermal load in °C*min
CTL_{ad}	Adjusted cumulated thermal load in K*s*10^5
d	Diameter in mm
ΔE	Total colour difference
g	Weight given to temperature value
L*	Lightness in the L*a*b* colour space
RH	Relative humidity in %
s	Slice thickness in mm
t	Time in s
T	Temperature in °C
v	Air velocity in m/s
z_C	Thermal destruction rate (for microbial inactivation)
μ	Mean or expectation of the distribution
σ^2	Variance

Subscripts

ad	Adjusted
air	Air
DP	Dew point
in	Inner
n	Number of measurement at point of time 0
n-1	Number of measurement at point of time -1
out	Outer
ref	Reference
S	Surface

Acknowledgements
The authors wish to acknowledge the Federal Ministry of Education and Research and the Federal Ministry for Economic Cooperation and Development for their financial support through the RELOAD GlobE project.

References

[1] A.-M. Nunez Vega, Simulation von Trocknungsprozessen empfindlicher biologischer Güter unter Berücksichtigung instationärer Randbedingungen, Dissertation, Kassel University (2015)
[2] D. Holdsworth, R. Simpson, Thermal Processing of Packaged Food, 2nd Edition, Springer Science and Business Media (2007)
[3] A.-M. Nunez Vega, S. Arendt, W. Speckle, W. Hofacker., Introducing a New Reference Value to Improve the Comparability of Quality Characteristics of Air Dried Foodstuffs, Conference Paper at 1st Nordic Baltic Drying Conference, Gdansk, Poland, (2015)
[4] A.-M. Nunez Vega, S. Arendt, W. Speckle, W. Hofacker, Cumulated Thermal Load as a Characteristic Value for Quality Changes of Air Dried Food, Conference Paper at 3rd North and East European Congress of Food (NEEFood), Brasov, Romania, (2015).
[5] S. Arendt, W. Hofacker, Post-Harvest Technology: How Thermal Processes Affect Product Quality, Conference Paper at 2nd (EN 14130) International Congress of Mechanical Engineering and Agricultural Science (CIIMCA 2015), Bucaramanga, Colombia, (2015)
[6] K. Joedicke, S. Arendt, W. Speckle, W. Hofacker., The Influence of Process Parameters on the Quality of Dried Agricultural Products Determined using the Cumulated Thermal Load, Conference Paper at EuroDrying, Liège, Belgium, (2017)
[7] R-BIOPHARM AG (Roche) Sucrose/D-Glucose/D-Fructose UV method for the determination of sucrose, D-glucose and D-fructose in foodstuffs and other materials
[8] EUROPEAN STANDARD EN 14130, Lebensmittel - Bestimmung von Vitamin C mit HPLC, Brüssel, (2003).

IOP Conf. Series: Materials Science and Engineering **437** (2018) 012002 doi:10.1088/1757-899X/437/1/012002

Characterization of three different solid wastes as energy resources for pyrolysis

F Campuzano[1], C F Ortíz[1], M Betancur[1] and J D Martínez[1]

[1] Grupo de Investigaciones Ambientales, Universidad Pontificia Bolivariana, Circular 1ra No. 70-01 Bloque 11, Medellín, Colombia

felipe.campuzano@upb.edu.co

Abstract. This work aimed the chemical and thermochemical characterization of three different solid wastes in order to identify promising properties for their further energy valorization by pyrolysis. The chosen materials were: (i) rice husk (RH), (ii) waste tire (WT), and (iii) municipal solid waste (MSW). The characterization techniques used were elemental, proximate, and heating value analyses. In addition, thermogravimetric analyses (TGA) in N_2 atmosphere were also conducted. The obtained results showed that RH has a higher heating value (HHV) of 15.30 MJ/kg and contains less carbon (~41 wt.%) and more oxygen (~38 wt.%) than conventional fossil fuels. The WT has a significant HHV (37.92 MJ/kg) mainly due to their high carbon content (~81 wt.%) and the low oxygen content (~2 wt.%). The MSW, free of metal and glass materials, was mainly composed of volatile matter (~92 wt.%), and showed a HHV of 22.41 MJ/kg. The TG diagrams exhibited different reaction zones which were associated with the decomposition of the main constituents of each sample. This simple but important characterization procedure could be considered as a fundamental tool for the qualification (expected characteristics of the final pyrolysis products) and quantification (maximum conversion efficiency) of the effect of some feedstock properties on the pyrolysis process. In this sense, this work is the starting point for the evaluation of pyrolysis process as an energy valorization method of different wastes.

1. Introduction

It is evident the current environmental and social concerns related to the management, valorization and disposal of those materials that have ended their life cycle. The final disposition of these materials, particularly those which present management challenges such as plastics and rubber, has been continuously regulated by a restrictive legislation that makes reuse, recycling and recovery procedures mandatory. These issues have forced the search of proper alternatives and pathways not only to be in line with the current normativity, but also to minimize environmental and public health drawbacks. Moreover, the great dependency on fossil fuels, especially in the transportation and power generation sectors, as well as the environmental impacts associated with their uncontrolled use have motivated some of the main research and development efforts up to date. This has been reflected in the increasing investment on renewable energy projects worldwide, which have been supported through a wide variety of international programs, especially those engaged toward the carbon market, such as the Clean Development Mechanism (CDM) [1], the Verified Carbon Standard [2], and the World Bank BioCarbon Fund [3], among others. In general, there are several factors that have driven the attention toward waste valorization including the environmental damaged case by unproperly disposal procedures, non or difficult biodegradability (*i.e.* plastics and rubber), the energy and material recovery potential (*waste to energy* processes) and the increasing need of recycling hydrocarbons to supply the energy demand for the upcoming years as a consequence of the fossil fuel depletion.

In Colombia, there are 1129 municipalities that produce around 28800 tons of solid wastes per day, with an annual growing rate between 4 and 8% [4]. It is estimated that a large fraction of these wastes is disposed inadequately in open dumps, rivers, lakes, and other environmental unsustainable fields, causing a devastating environmental impact. In addition, 110000 tons of waste tires (WTs) are produced each year becoming a serious problem since their very complex structure and composition (*i.e.* vulcanized rubber cured by sulfur, carbon black and silica) make difficult their post-treatment [5]. Tires are designed to be extremely resistant to physical, chemical and biological degradation and cannot be treated through simple processes. Moreover, their storage represents a big risk of fire that causes difficulties for their extinction with the subsequent production of highly toxic compounds leading to environmental problems in air, water, and soil, as occurred in Bogotá (Colombia) in 2014 [6] and Madrid (Spain) in 2015 [7]. Roughly speaking, waste management constitutes one of the most challenging problems of daily life in all the countries around the world. Waste management effectiveness depends on several aspects such as the level of technological development, the characteristics of waste, the government support by the establishment of adequate policies, and even the cultural level of the local population (recycling culture) [8].

Despite these drawbacks, municipal solid waste (MSW) and residual biomass have increased their importance as renewable source not only for their potential to produce heat and power using boilers and heaters, but also their ability to be transformed into more versatile fuels such as liquid ones, as well as different types of products for the chemical industry. Currently, biomass and wastes represent around 11 % of the total primary energy supply in the world, estimated to be about 13790 Mtoe for 2015 [9]. Although renewable sources are defined as those that can be replenished by nature (hydropower, wind power, solar power, etc.), the Environmental Protection Agency (EPA) has included MSW as a renewable energy resource pointing out that the energy contained in MSW is biogenic, especially when it comes from paper and organic wastes [10]. Similarly, WTs could be also considered as a renewable resource since they are produced, in some extent, from natural rubber which is obtained from the *Hevea* tree [11].

Moreover, it is important to highlight that these feedstocks do not threaten food supply and hence, it is not the source of “food or fuel” controversy [12]. It has been also stablished that energy production from residual biomass and wastes will offer to developing countries new income sources by creating new employment opportunities in new bioenergy infrastructures. Thus, this could help to reduce the rural poverty which is one of the Millennium Development Goals [13]. In this regard, bioenergy production from residual biomass and wastes seems to be one of the key options for the gradually substitution of fossil fuels (for instance as a counterpart), the mitigation of greenhouse gas emissions and the development of a more sustainable economy model by applying the concept of circular economy. Furthermore, these energy sources can deal with the concern of energy storage to address the problems associated with the intermittent production based on renewable energy such as wind and solar sources.

Power generation from residual biomass and MSWs can follow different conversion routes, known as *waste to energy* processes. These processes include thermochemical (combustion, gasification and pyrolysis) and biochemical (fermentation and bio-digestion) pathways which lead to liquid, solid and gas fractions with high added value in a wide range of applications. Nevertheless, the liquid fraction is particularly more interesting as an energy source because has higher energy density (in volume basis), as compared with gases and solids. Similarly, it can be easily storable, transportable and hence, distributable. In this sense, among all the conversion processes, several authors have highlighted that pyrolysis is much attractive due mainly to its less environmental impact and the possibility to recover liquid and solid compounds with wide potential in the energy and material sectors [14-16]. Therefore, it is considered that the energy contained in any feedstock is recovered in a cleaner way in contrast to other processes such as combustion and gasification. In pyrolysis, lower amounts of nitrogen oxides (NO_x) and sulfur oxides (SO_2) are produced due to the inert atmosphere and the lower operating temperatures. Even so, despite these advantages, the selection of any transformation pathway must be based on a detailed understanding of the feedstock physical-chemical properties and the goal fractions to be achieved (liquid, solid or gas).

Therefore, it is important to identify the main characteristics of the feedstock, including residual moisture (RM), volatile matter (VM), fixed carbon (FC), and ash content (proximate analysis), as well as its elemental composition (CHONS). Likewise, thermogravimetric analysis (TGA) provides important and useful information for determining the operational window of these feedstocks under pyrolysis conditions. This characterization is basic but fundamental to define not only proper transformation routes, but also operating conditions. It serves to estimate both maximum process efficiency and expected properties of the resulting pyrolysis products. Taking this into account, this work shows the main chemical and thermochemical properties of three solid wastes to assess their suitability to be considered as energy resources for pyrolysis. These wastes were selected since they are a good representation of the solid wastes generated in Colombia.

2. Materials and methods

The chemical and thermochemical analyses used in this work include elemental, proximate and heating value analyses and degradation patterns obtained from TGA. At first glance, this information allows to identify chemical properties, maximum conversion efficiencies and preliminary characteristics of the final pyrolysis products. In addition, TGA helps to define proper operating conditions, especially operation temperatures, in order to reach the maximum release of volatile matter when the feedstock is transformed through pyrolysis. These samples consisted of: (i) rice husk (RH) originally from the department of Tolima (Colombia), (ii) WT, supplied by a Colombian recycling company, and (iii) municipal solid waste (MSW), free of metal and glass materials, generated in one week at Pontificia Bolivariana University (UPB).

2.1 Sample preparation

First, RH sample was ground with a grinding machine to decrease particle size (up to 600 μm, or less). Second, 1.5 kg of a non-specified mixture of milled WT, without both the steel thread and the textile netting, was classified choosing particle sizes under 600 μm. Finally, the MSW sample was obtained from the solid wastes produced in one week at UPB. A mix of plastics, paper/paperboard and organics was selected to carry out all the analyses. These materials are the most abundant combustible fractions of MSW and are considered appropriated for being transformed through pyrolysis. The MSW sample was dried at 105°C for 2 hours in a muffle furnace Terrigeno model D18-Serie 1546, in order to avoid further decomposition and putrefaction. Later on, each material (plastics, paper/paperboard, and organic waste) was ground to decrease particle size.

2.2 Methods

2.2.1 Elemental, proximate and heating value analyses. The standard test methods used to conduct elemental analysis are shown in Table 1. Likewise, Table 2 reports the standard methods followed to perform the proximate and heating value analyses depending on the feedstock characteristics as well. The RM content was measured using a balance Shimadzu MOC-120H, while ash and VM content were determined gravimetrically by using a Muffle furnace Terrigeno Model D18-Serie 1546.

Table 1. Elemental analysis procedure

Parameter	Standard method	Equipment
C, H, N	UNE-EN 15407	Thermo Scientific
O	ASTM D5622-95 (by difference)	Elemental Analyzer
S	Application note 42151 Thermo Scientific	Flash 2000

Table 2. Proximate and heating value analyses procedures

Proximate analysis					Heating value
Feedstock	RM	VM	Ash	FC	
RH	UNE-CEN/TS 14774-3 ex	ISO-5623-1974	UNE-CEN/TS14775 ex	By difference	
WT	UNE-CEN/TS 14774-3 ex	UNE 32-019-84	UNE 32-004-84	By difference	ASTM 240-09
MSW	UNE-CEN/TS 14774-3 ex	ISO-5623-1974	UNE-CEN/TS14775 ex	By difference	

2.2.2 Thermogravimetric analysis (TGA). The TGA was conducted in a TA Q50 Thermal Analysis equipment with a weighing precision of ± 0.01 %. The TGA tests were performed for each waste at three different heating rates (5, 10 and 20°C/min) by using a N_2 volumetric flow of 100 ml/min. These samples were heated from room temperature (~25 ℃) to 700 °C and were kept at isothermal conditions for 20 minutes. The particle size used for these tests and all feedstocks was around 300 μm in order to assess only the chemical phenomena. The mass amount was between 14 and 25 mg depending on the material density. With the information collected through these analyses, the conversion grade (X) was determined by means of equation 1.

$$X = \frac{w_o - w_i}{w_o} . 100 \quad (\%) \tag{1}$$

Herein, w_o is the initial weight of the sample and w_i is the weight of the sample in each instant of time. TGA and DTG (derivative of TGA) feature important thermochemical parameters such as the initial degradation temperature (T_i) and the temperature where the maximum devolatilization rate (T_{max}) takes place. These parameters must be taken into account to understand the pyrolysis process of any feedstock. For instance, reactivity can be estimated by analyzing T_i and T_{max} in the DTG. Generally speaking, low T_i values and short periods of time to achieve it, account for more reactive materials since the VM starts being released easily. Likewise, low temperature levels to reach the T_{max} are also associated with high reactivity. From these parameters, it is possible to find a reactivity indicator (R_a) by means of equation 2, which is related to the maximum devolatilization peak in the DTG [17].

$$R_a = \left(\frac{1}{w_i}\right)\left(\frac{dw}{dt}\right)_{max} \quad (min^{-1}) \tag{2}$$

Where $(dw/dt)_{max}$ is the maximum weight loss rate (mg/min) in the DTG diagram. High R_a values are linked to high reactivity. This procedure was described by Ghetti [18] to study the reactivity of different coals.

3. Results and discussion

3.1 Rice husk

RH is an agricultural by-product with low density, which represents approximately 20% of the gross weight of the rice grain [19]. The RH represents a production globally around 822 million tons annually [20]. Generally speaking, RH is a complex mixture of three polymers: (i) hemicellulose $(C_5H_8O_4)_n$, where n=50–200, (ii) cellulose $(C_6H_{10}O_5)_n$, where n=500–4000, and (iii) lignin (mononuclear aromatic polymers), as well as extractives and ashes. Hemicellulose and cellulose are formed by a polymer chain shorter than the lignin one. From an energy point of view, lignin is considered one of the determining factors a long with cellulose for identifying the RH suitability for subsequent processing as an energy resource (production of bio-char, ethanol, etc.) [21]. Indeed, lignin exhibits a high correlation with the higher heating value (HHV) of lignocellulosic feedstocks, as shown in equation 3 [21], where *Lig* is the lignin content (wt.% dry basis).

$$\mathrm{HHV} = 87.7\,(\mathrm{Lig}) + 16495 \quad \left(\frac{\mathrm{kJ}}{\mathrm{kg}}\right) \tag{3}$$

Hemicellulose, cellulose and lignin varied depending on several factors including weather conditions, agronomic handling, type of soil and testing procedures, among others. For RH, hemicellulose ranges from 14 to 28.6 wt.%, while cellulose and lignin varied from 28.6 to 41.5 wt.% and from 20.4 to 33.7 wt.%, respectively [22]. In the Colombian case, Marrugo *et al.* [23] found that the RH has a hemicellulose, cellulose and lignin content of 24.50 wt.%, 39.65 wt.% and 35.84 wt.%, respectively (in dry, ash and extractives free basis). Table 3 shows the elemental and proximate analyses, as well as the heating value of the RH sample selected for this work.

Table 3. Elemental, proximate and heating value analyses of RH

Elemental analysis	wt.% (db)	Proximate analysis	wt.% (ar)	wt.% (db)
Carbon	41.13	VM	67.69	71.27
Hydrogen	3.37	RM	5.03	---
Nitrogen	0.33	Ash	16.15	17.00
Oxygen	38.17	FC	11.13	11.72
Sulfur	n.d.	Total	100.00	100.00
Total	100.00	HHV (MJ/kg)	14.53	15.30

n.d.: not detectable

As observed in Table 4, it is noted that RH contains less carbon (41.13 wt.%) and more oxygen (38.17 wt.%) than other fossil fuels such coal and petroleum-based fuels. Also, the HHV is lower than those fossil fuels (15.30 MJ/kg in dry basis). The high oxygen content could be considered as the main drawback when transforming RH by means of pyrolysis. For instance, the liquid fraction resulted from pyrolysis, will have acidic compounds leading to phase separation, repolymerization, coke formation, and insolubility when is blended to conventional fossil liquid fuels [24]. Therefore, it will show some difficulties for being used in most of the applications originally developed for conventional fuels. Nonetheless, several techniques such as hydrodeoxygenation and catalytic cracking, blending and emulsification, hydrocracking/steam reforming processes are currently under development in order to improve the quality of this liquid fraction for further implementation in conventional devices like internal combustion engines, boilers, etc. [24]. On the other hand, the RH biochar, the solid fraction resulting from RH pyrolysis, has demonstrated a high potential to offset carbon emissions by long term carbon sequestration with additional agricultural benefits such as improvements of soil qualities (fertilizers), and better plant growth [25].

The thermal decomposition of RH depicted in Figure 1 shows three different zones which are linked to moisture, hemicellulose, cellulose and lignin. The first zone is related to the moisture release, which takes place in a temperature range between 25 and 100°C. The second zone corresponds to hemicellulose-cellulose degradation at temperatures between 200 and 400°C. Finally, the third reaction zone, between 400 and 700°C, is mainly associated with lignin decomposition [26]. As observed, the degradation rate is no longer significant when temperature reaches 550°C, as most volatiles has already been released remaining the solid fraction (biochar). Moreover, T_i and T_{max} are found at ~210°C and ~370°C which are associated with the initial degradation of hemicellulose and the maximum degradation peak of cellulose, respectively. Herein, as the heating rate is increased, T_i is shifted to lower temperatures, while T_{max} is shifted to higher temperatures. This phenomenon is explained in detailed later. These results are in accordance to those reported in scientific literature [27].

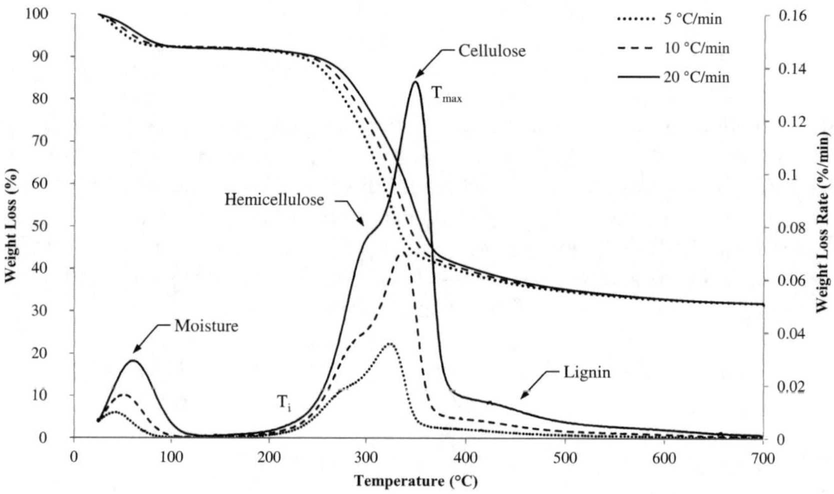

Figure 1. TGA and DTG of RH

Commonly, the behavior of lignocellulosic biomasses during pyrolysis has been referred to the performance of its natural components (hemicellulose, cellulose and lignin). In this sense, it is well-known the slow conversion of lignin, between 160-900°C, contrary to the faster decomposition of hemicellulose (220-315°C) and cellulose (315-400°C) [28]. However, these components may interact in different and complex manners along the process which suggests that the final pyrolysis products cannot be associated with the single composition and behavior of each constituent. In other words, when lignocellulosic biomasses are fed into a pyrolysis process, the interactions between these components show a significant influence on the formation of different chemical compounds contained in the liquid, the solid and the gaseous fraction [29].

3.2 Waste tire

Tires are made of rubber (both synthetic and natural rubber) (60-65 wt.%), carbon black (CB) (25-35 wt.%) and some accelerators and fillers which are added during the manufacturing process. Natural rubber (NR) comes from the *Hevea* tree, while synthetic rubber (butyl rubber (BR) and styrene-butadiene (SBR) copolymer) is generally derived from petroleum-based products. CB, an amorphous carbon of quasi-graphitic structure, is produced mainly by partial combustion of fossil hydrocarbons. Finally, fillers and accelerators consist of organic and inorganic compounds [30]. This composition changes depending on brand, age and type of tire (passenger car tire (PCT) or truck tire (TT)). Table 4 points out both the elemental and proximate analyses, as well as the heating value of the WT sample selected for this work.

Table 4. Elemental, proximate and heating value analyses of WTs

Elemental analysis	wt.% (db)	Proximate analysis	wt.% (ar)	wt.% (db)
Carbon	81.49	VM	60.66	60.96
Hydrogen	8.21	RM	0.50	---
Nitrogen	< 0.10	Ash	6.48	6.51
Oxygen	1.74	FC	32.35	32.51
Sulfur	2.05	Total	100.00	100.00
Total	100.00	HHV (MJ/kg)	37.73	37.92

As expected, WT have a significant HHV of 37.92 MJ/kg (in dry basis) which is greater than that of coal (bituminous coal which range from 24.5 to 32.6 MJ/kg [17]). This is due to the higher content of carbon (81.49 wt.%). Likewise, the low amount of oxygen (1.74 wt.%) and moisture (0.5 wt.%) lead to the formation of a high-quality liquid fraction when this feedstock is pyrolyzed, with almost neutral pH, stability, and moderated viscosity and density, among others. Although it can exhibit up to 1 wt.% of sulfur, which hinders its direct application as an alternative fuel [31], this liquid fraction is considered as a potential source in the petrochemical market [32]. Several upgrading techniques are currently being studied to improve these properties [33]. Similarly, the CB added during the manufacturing tire process is also recuperated after pyrolysis. The CB as well as the inorganic compounds are trapped in the solid fraction [34]. Among others, this carbonaceous material can be used for activated carbon production [34], or as solid fuel (HHV ~30 MJ/kg). However, the most attractive application is as substitute of commercial CB, after being upgraded/refined [35].

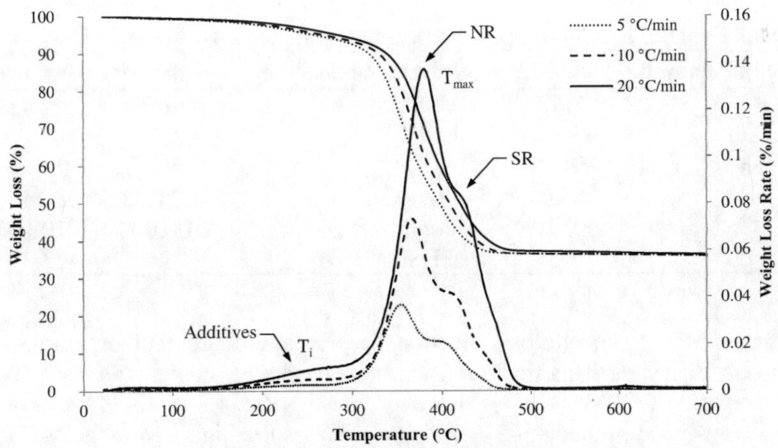

Figure 2. TGA and DTG of WT

The TGA and DTG for WTs also present three different zones related to the degradation of rubber compounds (NR, SBR and BR), as can be noticed in Figure 2. The first zone (200-300°C) is related to the tire rubber additives decomposition; the second zone (310-450°C) (highest peak) corresponds to the NR decomposition; and finally, the third zone (400-500°C) is linked to the synthetic rubber (BR and SBR) degradation [36]. Furthermore, T_i is reached at ~230 °C and is associated with the additives devolatilization, while T_{max} is obtained at ~370°C and is linked to the NR degradation. On the other hand, as the heating rate increases, the weight loss profile is shifted to higher temperature, as reported elsewhere [37]. This means that the degradation rate increases as the heating rate is increased. Likewise, the temperature at which the maximum devolatilization begins and ends is also affected. This phenomenon is attributed to the combined effects of the heat transfer and the changes of devolatilization kinetics of each component [38], resulting in delayed decomposition in terms of temperature. Finally, it is worth to note that pyrolysis reactions start at 200 °C and finish at 480 °C, approximately. Therefore, lower temperatures than 500°C could lead to the incomplete conversion of the WT, especially at high particle sizes.

3.3 Municipal solid waste

As commented before, the MSW used in this work is a mixture of plastics, paper/paperboard and organic matter. The plastic waste consisted of plastic battles (from water, soda, oil, juice, etc.), as well as plastic bags and food packages, among others. It considers a mixture of polyethylene (PE) (low density (LDPE) and high density (HDPE)), polypropylene (PP), and polystyrene (PS). The polymeric structure of both LDPE and HDPE is essentially a long chain of aliphatic hydrocarbons. The difference in density is due to the chain branching [39, 40]. In contrast, PP has a different structure than both LDPE and HDPE given the presence of a methyl group (CH_3), while PS is made of styrene monomer and contains a benzene ring (C_6H_6) [40]. On the other hand, paper/paperboard waste comes from newspaper, boxes, glossy paper, cardboard, etc. The common element of these materials is pulp obtained from wood. Wood is composed by cellulose (40-45 wt.%), the skeletal polysaccharide; hemicellulose (27-39 wt.%) which form the matrix; and lignin (21-30 wt.%), the encrusting substance that bins the cells together [40]. Additionally, wood contains many low molecular-weight organic compounds (2-5 wt.%) and small amount of mineral matter (0-3 wt.%). Finally, the organic waste is a mix of different fruit peels and vegetable wastes. All they are mainly lignocellulosic materials and the content of cellulose, hemicellulose and lignin depends on their specific nature. Table 5 shows the elemental and proximate analyses, as well as the heating value of the MSW. It is important to point out that the RM reported in Table 5 is the final moisture once the sample was dried and stabilized to avoid putrefaction and decomposition (initially with a moisture content around 80 wt.%).

Table 5. Elemental, proximate and heating analyses of the MSW

Elemental analysis	wt.% (db)	Proximate analysis	wt.% (ar)	wt.% (db)
Carbon	55.67	VM	87.69	92.45
Hydrogen	6.73	RM	5.15	---
Nitrogen	< 0.10	Ash	2.28	2.40
Oxygen	35.20	FC	4.88	5.14
Sulfur	< 0.10	Total	100.00	100.00
Total	100.00	HHV (MJ/kg)	21.26	22.41

The properties of the lignocellulosic fraction (paper/paperboard and organic wastes) are quite different in contrast to plastics. This fraction usually presents a FC content between 10-20 wt.%, while the VM ranges between 20-30 wt.% [41]. However, plastics contain up to 90 wt.% of VM since they come from petroleum-based products [42]. This fact explains the high VM in the MSW (92.45 wt.% in dry basis) as well as the HHV (22.41 MJ/kg in dry basis) due to the high share of plastics (up to 50 wt.%) into the MSW selected in this work. Similar to RH, the main drawback of the MSW sample for being transformed by means of pyrolysis is the high oxygen content (35.2 wt.%).

On the other hand, Figure 3 shows the TGA and DTG of the MSW. Similar to the results obtained for RH and WTs, different zones can be identified. However, as the sample is a complex mixture of different components, the identification of each compound, as made for the other samples, represents a difficult task. However, Sorum *et al.* [40] studied the plastics and paper behavior separately by using TGA/DTG in order to have a better understanding of this phenomenon. They found that the major weight loss of cellulosic matter (paper) occurred at 250-400 °C, with a maximum devolatilization point between 355 and 371 °C. These results were expected taking into account the negligible presence of lignin in the feedstock analyzed by authors which reduces the degradation time. On the other hand, the degradation of plastics (PS, PE and PP) presented a slight weight loss within a narrow temperature range from 80 to 100 °C. Then, the major devolatilization took place at temperatures of 350 and 500 °C with a maximum peak between 413 and 479 °C. Similarly, Miranda *et al.* [43] analyzed the thermal decomposition of sweet orange dry peel by using TGA/DTG at different heating rates (1, 5, 10, 20, and 40 °C/min). They found four stages of weight loss and four peaks at 47 °C, 223 °C, 346 °C, and 479 °C, which were associated with degradation of water soluble components, hemicellulose, cellulose and lignin, respectively. According to these results, it can be inferred that the first peak in Figure 3, between 25 and 100 °C, can be attributed to the release of moisture and some light hydrocarbons (LH) (coming from plastics which degrade in a narrow temperature range of 80-100 °C). The second and third peak can be related to degradation of hemicellulose/cellulosic compounds (paper, organic waste). The fourth peak is linked to the maximum weight loss rate of plastics; while the fifth peak is associated with the lignin degradation.

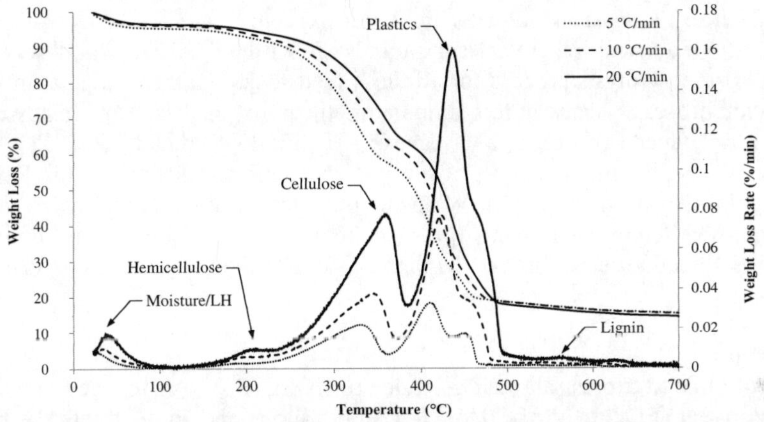

Figure 3. TGA and DTG of MSW

Finally, Table 6 gathers the T_i, the T_{max} and the devolatilization velocity associated with the highest peak (*Ra*) for all the samples. Likewise, Table 7 reports the (C+H)/O and VM/FC ratios, which are considered a chemical criterion to evaluate the reactivity, especially in the case of gasification and/or combustion applications [17]. According to Ghetti [18], feedstock with high oxidation grade, which means low (C+H)/O ratio, are more reactive during combustion and gasification. Similarly, and under oxidations conditions, the higher the VM/FC ratio, the higher the devolatilization rates and also a more reactive feedstock [17]. In the case of pyrolysis process, it is difficult to determine the reactivity from these relations because more complex kinetic phenomena are involved. Even so, the VM/FC ratio provides the theoretical process efficiency in terms of volatiles and solid fraction yields.

Table 6. Reactivity analysis

Feedstock	Heating rate (°C/min)	T_i (°C)	T_{max} (°C)	(C+H)/O	VM/FC	R_a (min^{-1})
	5	210	321			0.036
RH	10	208	334	1.26	6.08	0.070
	20	205	347			0.134
	5	300	348			0.037
WT	10	294	362	29.31	1.86	0.074
	20	212	375			0.138
	5	267	407			0.034
MSW	10	246	422	1.94	17.96	0.079
	20	194	435			0.162

As observed RH and MSW are more reactive for combustion and/or gasification applications in contrast to WTs due to their lower oxidizing grade (lower (C+H)/O ratio). However, RH and MSW tend to release more VM than WT due to their higher ratios of VM/FC, which could be benefit for pyrolysis for obtaining more liquid and gas fractions. In terms of degradation temperature and velocity (R_a), it can be inferred that these values are strongly influenced by the heating rate. Therefore, it is seen that higher heating rates lead to higher devolatilization velocities and T_{max}, but lower T_i.

4. Conclusions

RH, WT and MSW were analyzed by elemental, proximate and heating value analyses. WT has the greater HHV (37.92 MJ/kg in dry basis), followed by MSW (22.41 MJ/kg in dry basis) and RH (15.30 MJ/kg in dry basis). Also, WT can be regarded as a promising feedstock for pyrolysis mainly due to the low oxygen (~2 wt.%) but high carbon (~81 wt.%) that promote the formation of a high-quality liquid fraction to be used as a fuel. Contrary, RH and MSW are characterized for a high oxygen content (~38 and ~35 wt.%, respectively) which lead to the formation of oxygenated compounds in the liquid fraction after pyrolysis. Hence, a low pH and high instability are expected. However, they can serve as precursors for interesting and value-added chemical compounds. From the TGA, it was observed that 550°C is enough to conduct the pyrolysis process for all these feedstocks. Higher temperatures do not improve the devolatilization process because at this temperature the pyrolysis has been already completed (in the case of WT) or is no longer significant (in the case of RH and MSW). In addition, DTG showed different degradation zones for all the components of the WT (NR, BR and SBR) and the RH (hemicellulose, cellulose and lignin). However, for the MSW this identification is more complicated since the sample is complex mixture of different components (plastics, hemicellulose, cellulose and lignin). On the other hand, RH and MSW tend to release more VM than WT due to the higher ratios of VM/FC (~ 6 and ~18 respectively).

Acknowledgments

The authors would like to express all their gratitude to COLCIENCIAS for the financial support given by the research project (FP44842-281-2016). F. Campuzano is also in debt with the Integrated Center for the Development of Research (CIDI) at UPB and its program *"Formación Investigativa"* for the MSc scholarship.

References

[1] Office of Energy Efficiency and Renewable Energy. Bioenergy Technologies Office. [Internet]. Available: http://energy.gov/eere/transportation/bioenergy. [Cited: 06-Apr-2016].
[2] Australian Renewable Energy Agency, Bioenergy projects. [Internet]. Available: http://arena.gov.au/projects/bioenergy/. [Cited: 06-Apr-2016].
[3] Lee CM, Lazarus M. Bioenergy projects and sustainable development: which project types offer the greatest benefits?. Climate and Development. 2013; 5(4): 305-317.
[4] Correal MC. Solid waste management in Colombia. Waste Management for Everyone. [Internet]. Available: http://www.d-waste.com/new-infographics/item/114-solid-waste-management-in-colombia.html., 2012.
[5] Sandoval MA. Argos utilizará las llantas usadas como combustible. El Colombiano. [Internet]. Available: http://www.elcolombiano.com/antioquia/argos-utilizara-las-llantas-usadas-como-combustible-HE1251832., 2015 [Cited: 05-Sep-2016].
[6] Penagos JG. Lo que dejó al descubierto el incendio en Fontibón. El Espectdor. [Internet]. Available: http://www.elespectador.com/noticias/bogota/dejo-al-descubierto-el-incendio-fontibon-articulo-527416., 2014 [Cited: 05-Sep-2016].
[7] Barroso FJ, León P. Un gran incendio en Seseña arrasa el mayor cementerio de neumáticos de España. El País. [Interet]. Available: https://elpais.com/ccaa/2016/05/13/madrid/1463120336_194139.html.m., 2015 [Cited 22-Jun-2017].

[8] Barba D, Brandani F, Capocelli M, Luberti M, and Zizza A. Process analysis of an industrial waste-to-energy plant: Theory and experiments. Process Safety and Environmental Protection. 2015; 96: 61–73.

[9] International Energy Agency. Key world energy statistics. [Internet]. Available: http://www.iea.org., 2017 [Cited: 14-Apr-2018].

[10] Cheng H, Hu Y. Municipal solid waste (MSW) as a renewable source of energy: current and future practice in China. Bioresource Technology. 2010; 101: 3816-3824.

[11] Shulman VL. Tyre Recycling. European Tyre Recycling Association (ETRA). [Internet]. Available: http://www.polymerjournals.com/pdfdownload/928289.pdf. [Cited: 14-Apr-2017].

[12] Thompson PB. The agricultural ethics of biofuels: the food vs. fuel debate. Agriculture. 2012; **2**: 339-358

[13] Sachs JD, McArthur JW. The Millennium Project : a plan for meeting the Millennium Development Goals. Lancet. 2015; 365: 347–353.

[14] Aylón E, Fernández-Colino A, Murillo R, Navarro MV, García T, Mastral AM. Valorisation of waste tyre by pyrolysis in a moving bed reactor. Waste Management. 2010; 30: 220–1224.

[15] Murillo R, Aylón E, Navarro MV, Callén MS, Aranda A, Mastral AM. The application of thermal processes to valorise waste tyre. Fuel Processing Technology. 2006; 87: 143–147.

[16] Williams PT. Pyrolysis of waste tyres: a review. Waste Management. 2013; 33: 1714–1728.

[17] Perez JF, Barrera R, Salazar C. Producción de gas natural sutético mediante gasificación de carbones Colombianos-análisis termodinámico en Aspen Plus. Universidad de Antioquia. 2015.

[18] Ghetti P. DTG combustion behavior of coal-correlation with proximate and ultimate analysis date. Fuel.1986; 65: 636–639.

[19] Ramirez JJ, Martínez JD. Performance analysis of an air-blown pilot fl uidized bed gasifier for rice husk. Energy for Sustainable Development. 2014; 18: 75–82.

[20] Dunnigan L, Ashman PJ, Zhang X, Kwong CW. Production of biochar from rice husk: Particulate emissions from the combustion of raw pyrolysis volatiles. Cleaner Production. 2018; 172: 1639-1645.

[21] Quispe I, Navia R, Kahhat R. Energy potential from rice husk through direct combustion and fast pyrolysis : a review. Waste Managament. 2017; 59: 200–210.

[22] Pottmaier D, Costa M, Farrow T, Oliveira AAM, Alarcon O, Snape C. Comparison of rice husk and wheat straw: from slow and fast pyrolysis to char combustion. Energy and Fuels. 2013; 27: 7115-7125.

[23] Marrugo G, Valdés CF, Chejne F. Characterization of Colombian agroindustrial biomass residues as energy resources. Energy and Fuels. 2016; 30: 8386-8398.

[24] Hossain AK, Davies PA. Pyrolysis liquids and gases as alternative fuels in internal combustion engines – A review. Renewable and Sustainable Energy Reviews. 2013; 21: 165–189.

[25] Abrishamkesh S, Gorji M, Asadi H, Bagheri-Marandi GH, Pourbabaee AA. Effects of rice husk biochar application on the properties of alkaline soil and lentil growth. Plant Soil Environment. 2015; 61: 475–482.

[26] Brown RC. Chapter 5. Fast Pyrolysis. Thermochemical Processing of Biomass-Conversion into Fuels, Chemical and Power. John Wiley and Sons.

[27] Alias N, Ibrahim N, Kamaruddin Abd Hamid M, Hasbullah H, Ali RR, Sadikin AN, Asli UA. Thermogravimetric analysis of rice husk and coconut pulp for potential biofuel production by flash pyrolysis. The Malaysian Journal of Analytical Science. 2014; 18 (3): 705–710.

[28] Yang H, Yan R, Chen H, Lee DH, Zheng C. Characteristics of hemicellulose , cellulose and lignin pyrolysis. Fuel. 2007; 86: 1781–1788.

[29] Kan T, Strezov V, Evans TJ. Lignocellulosic biomass pyrolysis : a review of product properties and effects of pyrolysis parameters. Renewable and Sustainable Energy Reviews. 2016; 57: 1126–1140.

[30] Martínez JD, Puy N, Murillo R, García T, Navarro MV, Mastral AM. Waste tyre pyrolysis: a review", Renewable and Sustainable Energy Reviews. 2013; 23: 179–213.

[31] Hu H, Fang Y, Liu H, Yu R, Luo G, Liu W, Li A, Yao H. The fate of sulfur during rapid pyrolysis of scrap tires. Chemosphere. 2014; 97:102-107.
[32] Murugan S, Ramaswamy MC, Nagarajan G, The use of tyre pyrolysis oil in diesel engines. Waste Management. 2008; 28 (12): 2743–2749.
[33] Hita I, Arabiourrutia M, Olazar M, Bilbao J, Arandes JM, Castaño P. Opportunities and barriers for producing high quality fuels from pyrolysis of scrap tires. Renewable and Sustainable Energy Reviews. 2016; 56: 745-759.
[34] Betancur M, Martínez JD, Murillo R. Production of activated carbon by waste tire thermochemical degradation with CO_2. Journal of Hazardous Materials. 2009; 168: 882–887.
[35] Martínez JD, Murillo R, García T. Production of carbon black from the waste tires pyrolysis. Grupo Español del Carbon. 2013; 30: 10-14.
[36] Martínez JD, Murillo R, García T, Veses A. Demonstration of the waste tire pyrolysis process on pilot scale in a continuous auger reactor. Journal of Hazardous Materials. 2013; 261: 637–645.
[37] Williams PT, Besler S. Pyrolysis-thermogravimetric tyres and tyre components analysis of tyres and tyre components. Fuel. 1995; 14: 1277–1283.
[38] Martínez JD. Waste tire valorization by pyrolysis : production and application of the products Ph.D. Thesis, Universidad de Zaragoza. 2014.
[39] Brydson JA. Plasctic materials. Butterworth-Heinemann. Seventh edition. 1999.
[40] Sørum L, Grønli MG, Hustad JE. Pyrolysis characteristics and kinetics of municipal solid wastes. Fuel. 2001; 80: 1217–1227.
[41] ECN Phyllis 2. Database for biomass and waste. [Internet]. Available: https://www.ecn.nl/phyllis2/., 2016. [Cited: 10-Sep-2016].
[42] Sharuddin SDA, Abnisa F, Daud WMAW, Aroua MK. A review on pyrolysis of plastic wastes. Energy Conversion Management. 2016; 115: 308–326.
[43] Miranda M, Bustos-Martínez D, Sosa C, Gutiérrez MH, Rodriguez ME. Pyrolysis of sweet orange (Citrus sinensis) dry peel. Journal of Analytical and Applied Pyrolysis. 2009; 86: 245–251.

IOP Conf. Series: Materials Science and Engineering **437** (2018) 012003 doi:10.1088/1757-899X/437/1/012003

Effect of temperature, starch and plasticizer concentrations on color parameters of ulluco (*Ullucus tuberosus* Caldas) edible films

A K Homez[1], L D Daza[1], J F Solanilla[2] and H A Váquiro[1]

[1]Centro de Desarrollo Agroindustrial del Tolima, CEDAGRITOL, Departamento de Producción y Sanidad Vegetal, Facultad Ingeniería Agronómica, Universidad del Tolima, 730006, Ibagué, Colombia
[2]Departamento de Agroindustria, Facultad de Ciencias Agrarias, Universidad del Cauca, 190003, Popayán, Colombia

havaquiro@ut.edu.co

Abstract. The growing concern about limited natural sources and pollution has aroused the interest of developing bio-packaging based on agricultural polymers. The Andean tuber ulluco (*Ullucus tuberosus* Caldas) is a promising source of starch that can be used in the production of edible films or coatings for preserving the quality of food products. Color is an important feature for edible films production due to can alter the visual perception of the coated product. In this study, the effect of drying temperature (*T*) (40 to 60°C), starch concentration (*SC*) (2.0 to 3.0% w/v) and glycerol concentration (*GC*) (1.0 to 3.0% v/v) on the color parameters (L*, a* and b*) of ulluco edible films was evaluated. A composite central design was used to combine five levels of the three factors (*T*, *SC* and *GC*). Artificial neural network (ANN) were applied to establish mathematical models that could represent the dependence of the factors on the color parameters. The edible films showed luminosity values between 87.4 and 90.3. Results indicated that the higher *T*, the lower a* and b* parameters and the higher luminosity of the films. Likewise, the lower *G*, the higher luminosity of edible films. Sensibility analysis shows that *SC* has an important influence on L*, a*, b* parameters. Statistical results indicated that the ANN approach provided an accurate estimation (R^2_{adj} = 0.885 and *MSE* = 0.054) of the color parameters. ANN use a feed-forward architecture with five neurons in the hidden layer, hyperbolic-tangent transfer function, and Bayesian regularization backpropagation as training algorithm. ANN could be an alternative tool for modelling the complex relationship between the color parameters and the temperature, starch and plasticizer concentrations on ulluco edible films.

1. Introduction

Over the last twenty years has been growing the global concern about pollution generated by the indiscriminate use of non-renewable products. As result, many researches are focused on the development of products that can use as food packaging and replace petroleum based materials. Bioplastics are an alternative solution for this issue, they are obtained from natural polymers and their demand has been increased, by 2019 their production is expected to be around 5.5 million metric tons [1].

Biodegradable films can be produced from different materials such as starch, cellulose derivates, chitosan, gums and animal or plant based proteins [2]. Among these materials, starch is one of the most promising due to their biodegradability, biocompatibility, low cost and wide availability in renewable sources. Starch-based films have been studied in a wide range at experimental level, and present suitable characteristics for their use as food packaging such as transparency, odorless and tasteless [3]. In addition, starch-based films can extend shelf life of food by acting as barrier against mechanical damages or controlling gases exchange [4].

Color properties of edible films are an important feature in packaging, since they can alter the visual perception of food products [5]. However, the performance of edible films based on starch may vary, this is related to its chemical structure, specifically to the amylose/amylopectin ratio, which depends on the botanical source of starch [6]. Basiak, Lenart, and Debeaufort [6] studied the effect of starch type on the color properties of edible films, found that the starch amylose content has a negative effect on transparency of wheat, corn and potato starch based films.

Ulluco (*Ullucus tuberosus* C.) is an Andean highland tuber that normally is cultivated between 2400 and 4250 meters above sea level. Ulluco tubers are small (2-15 cm) and have oblong or spherical shape, the color tuber skin may vary from yellow, orange, red, magenta to purple, while, the pulp is white or yellow [7]. Ulluco tuber is rich in carbohydrates, protein, fiber and vitamin C [8]. In Colombia, the tuber is consumed *in natura* and used in typical preparations. The use of the tuber is related to old indigenous traditions of paeces, muiscas and lanches from the Andean zones from Cauca, Nariño, Boyacá and Cundinamarca departments [9]. According to Cruz et al. [10] ulluco starch has suitable characteristics, such as high gel stability, for being use in food applications. However, there is not enough information about the possible industrial uses, making it difficult to take full advantage of this raw material. Therefore, the addition of value to ulluco would have a positive impact on the regional rural economy.

In this context, the aim of this work was to evaluate the effect of drying temperature (40°C to 60°C), ulluco starch concentration (2% to 3% w/v) and glycerol concentration (1% to 3% w/v) on the L*, a* and b* parameters of ulluco starch based edible films using artificial neural network (ANN).

2. Materials and Methods

2.1. Materials

Ulluco tubers were purchased in a local market in the city of Ibagué (Colombia). The samples were selected and washed, and those that showed mechanical damage on the surface were discarded.

2.2. Starch Extraction

Extraction of ulluco starch was performed according to the methodology describe by Valcárcel-Yamani et al. [11].

2.3. Experimental design

A central composite inscribed design (*CCI*) was used to determine the effect of drying temperature *T* (°C), starch concentration *SC* (% w/v) and glycerol concentration *GC* (% w/v), on the optical properties of ulluco edible film. Five levels of each factor were combined to obtain a 15 formulations design matrix (Table 1). Each formulation was prepared by duplicated and the experiments were carried out by triplicated, totalizing 90 experiments.

Table 1. Effect of *SC*, *GC* and *T* on optical properties of ulluco starch edible film

Sample	*SC* (%w/v)	*GC* (%w/v)	*T* (°C)	L*	a*	b*
1	2.2	1.4	44	88.8 ±0.57	-0.23 ±0.03	2.90 ±0.06
2	2.2	1.4	55	89.2 ±0.53	-0.31 ±0.02	2.95 ±0.06
3	2.2	2.6	44	87.5 ±0.27	-0.30 ±0.00	3.06 ±0.04
4	2.2	2.6	55	88.2 ±0.30	-0.36 ±0.02	3.28 ±0.06
5	2.8	1.4	44	89.2 ±0.19	-0.10 ±0.02	2.87 ±0.06
6	2.8	1.4	55	89.1 ±0.26	-0.04 ±0.02	2.75 ±0.03
7	2.8	2.6	44	88.3 ±0.27	-0.16 ±0.01	2.99 ±0.02
8	2.8	2.6	55	90.3 ±0.32	-0.22 ±0.01	3.03 ±0.10
9	2.0	2.0	50	88.8 ±0.24	-0.06 ±0.01	2.89 ±0.04
10	3.0	2.0	50	89.1 ±0.21	-0.24 ±0.01	2.99 ±0.04
11	2.5	1.0	50	88.8 ±0.37	-0.44 ±0.02	2.97 ±0.05
12	2.5	3.0	50	87.4 ±0.11	-0.26 ±0.02	3.02 ±0.03
13	2.5	2.0	40	87.6 ±0.16	-0.28 ±0.02	3.05 ±0.02
14	2.5	2.0	60	87.7 ±0.17	0.12 ±0.01	2.93 ±0.05
15	2.5	2.0	50	87.8 ±0.38	0.07 ±0.01	2.92 ±0.07

SC: starch concentration; *GC*: glycerol concentration and *T*: temperature

2.4. Film Preparation

Ulluco starch films were prepared by casting following the methodology described by Tumwesigye, Montañez, Oliveira, & Sousa-Gallagher [12], with some modifications. Starch at different concentrations (Table 1) was dissolved in distilled water using magnetic stirring for 5 min. The mixture was heated to 95°C and held at this temperature above 5 min, until a clear viscous gel was obtained. The mixture was cooled down to 50°C and then glycerol (Merck, Millipore, MA, USA) was added at different concentrations (Table 1). 30 g of solution were transferred to a Petri dish of 8.5 cm diameter. The samples were dried for 24 h in an oven with forced air at different temperatures (Table 1). Moisture films was stabilized in a desiccator cabinet (DH-1002, Acequilab, Colombia) at 54% RH for 24 h, prior to analyzes.

2.5. Color

Film color was determined by Minolta colorimeter (Cr 410, Konica Minolta, Japan) in the CieLab scale. Standard white plate was used to calibrate the equipment (Y=86.6, x=0.3153, y=0.3228). The L*, a*, and b* values were evaluated by reflectance measurements. In these system, L* is the luminosity (from black to white), and the horizontal axes a* and b* are the chromatic coordinates (from -a*: greenness, -b*: bluish to +a*: redness, +b*: yellowness) [5].

2.6. Artificial Neural Network (ANN) model

ANN was used to determine the effect of temperature (*T*), starch concentration (*SC*) and glycerol concentration (*GC*) on the L*, a* and b* parameters of ulluco edible films. A feed-forward neural network was developed and trained by Bayesian regularization backpropagation (trainbr) algorithm (Table 2). Transfer function log-sig was used. Three-layer basic architecture was modelling, the input layer with three neurons reflecting the independent variables (*T, SC, GC*), a hidden layer with five neurons and the output layer representing the dependent variables (L*, a*, b*). All the experimental data were used for ANN modeling. ANN training and simulation were carried out by the Neural Network Toolbox of Matlab® R2016b (The Mathworks Inc., Natick, MA, USA) software. Both input and output data were normalized within the range [-1,1] by the ''mapminmax'' function of Matlab® which helps to reduce network error by balancing the order of magnitude of the inputs [13].

ANN performance was evaluated by the adjusted coefficient of determination (R^2_{adj}) (Equation. 1) and the mean square error (*MSE*) (Equation 2).

$$R^2_{adj} = 1 - (1 - R^2)\frac{n-1}{n-m} \tag{1}$$

$$MSE = \frac{1}{n}\sum_{i=1}^{n}(x_i^* - x_i)^2 \tag{2}$$

where x_i^* represents the experimental values; x_i represents the estimated values; n is the number of experimental or estimated values and m is the number of estimated parameters (biases and weights).

The best ANN configuration was selected for L*, a* and b* parameters, first, according to the highest R^2_{adj} and the lowest *MSE*, and secondly, considering the lowest number of hidden-layer neurons and the simplest network architecture. R^2_{adj} evaluated how successful each ANN explained the variation in the experimental data by adjusting the coefficient of determination (R^2) based on the number of biases and weights [14].

Moreover, the selected ANN configurations were submitted to a residual analysis, to assess the adequacy of ANN models [15], and were submitted to a global sensitivity analysis by the Sobol' method [16], to quantify the effect of factors and their interactions on the responses.

Table 2. Parameters of artificial neural network (ANN)

Properties	ANN architecture	Transfer function	Neurons (hidden layer)	Number of parameters	R^2	R^2_{adj}	*MSE*
Color	Feed-forward	Logsig	5	38	0.901	0.885	0.054

ANN: artificial neural network and *MSE:* mean square error

3. Results and discussion

Edible films color is an important characteristic for consumers acceptability as packaging material. Films with high gloss and transparency leads to a good visual food presentation [3]. In general, ulluco starch edible films had good appearance and clarity. Ulluco films were homogenous and easily removed from the plastic petri dish. As exposed Figure 1, the film does not alter food natural colors, which shows good characteristics for being applied as food packaging.

The effect of *SC*, *GC* and *T* on the L*, a* and b* parameters are shown in the Figure 2. L* values ranged between 87.4 and 90.3 (Table 1). The highest L* value was shown by sample prepared with 2.8% of *SC*, 2.6% of *GC* and 55°C of *T*. The luminosity values obtained in this work were higher than those reported for babassu mesocarp starch edible films (L* = 76) [17] and mucilage-based films produced with *Cereus hildmANNian* (L* = 86) [18].

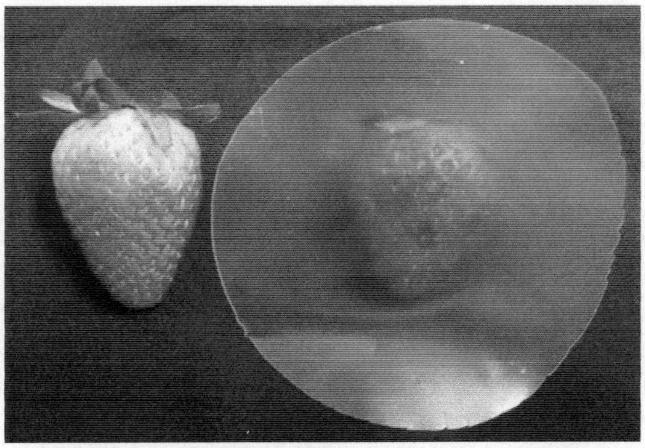

Figure 1. Ulluco edible films appearance

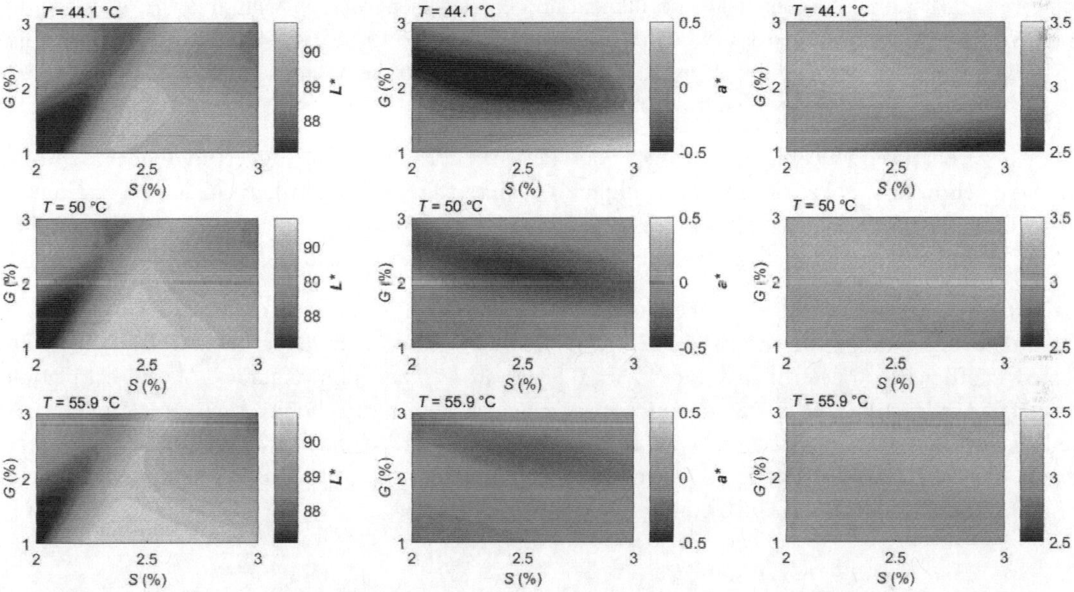

Figure 2. Effect of *SC*, *GC* and *T* on color properties of ulluco starch edible film. *S*: starch concentration; *G*: glycerol concentration and *T*: temperature

Sensitivity analysis (SA) allows the calculation of two factors *Si* and *STi* (Table 3), the first order sensitivity index (*Si*) measures only the main effect contribution of each input factor on the output variance. However, the second index (*STi*) shows the contribution of the main factor interactions between them and the rest [16]. Table 3 shows that *SC* was the factor with the greatest sensitivity on L* values, which is consistent with the values obtained of ulluco edible films at 2.8% and 3% of *SC* (Table 1). However, SA also shows that the interaction between *SC* and *GC* parameters have effect on L* values. Fig. 2 illustrate these results, when *SC* increased and *GC* decreased the luminosity of the edible films are higher, this could be explained since the increase in plasticizer concentration (*GC*) leads to a polymeric chain compaction that can modified the refractive index and restricting the passage of the light through the film matrix [19]. Moreover, *T* parameter does not show a sensitivity effect on luminosity.

Table 3. First-order-effect and total-effect sensitivity indices corresponding to SC, GC and T

Factor	*L**			*a**			*b**		
	S_i	ST_i	Difference	S_i	ST_i	Difference	S_i	ST_i	Difference
SC	0.45	0.877	0.427	0.108	0.341	0.233	0.102	0.249	0.147
GC	0.058	0.47	0.412	0.39	0.777	0.387	0.474	0.751	0.277
T	0.04	0.116	0.076	0.045	0.404	0.359	0.061	0.395	0.334

SC: starch concentration; *GC*: glycerol concentration; *T*: temperature; *Si*: sensitivity index and *STi*: total sensitivity index

SA shows that a* and b* parameters have similar behavior, both were mainly sensitive (*Si*) to *GC*, followed by *SC* and last by *T*. In addition, *GC* and *T* interactions (ST_i) have the greater sensibility effect on a* and b* parameters. a* values ranged between -0.36 to 0.12 (Table 1), showing that ulluco edible films tend to a neutral color. Fig. 2 shows when *GC* and *SC* increased a* values decreased and slightly tends to greenness. On the other hand, the use of middle drying *T* produces that a* values tend to zero. Nawab et al. [3] studied the physical, mechanical and barrier properties of Mango kernel starch-gum composite films, they observed lower a* values (-0.86) compare to those obtained for ulluco edible films, indicating a higher tendency to green. This can be related to the botanical source of raw material (starch) used in the preparation of edible films.

As can be observed in Table 1, b* values were between 2.75 and 3.28, suggesting that these films tend to the yellowness. It was observed that higher b* values were obtained when *GC* increased (Figure 2). However, the opposite behavior was observed for *SC* and *T* variables. The results obtained in this study for b* values were lower than those reported by Saberi et al. [19] and Vargas et al. [20].

ANN is a computational and mathematical modelling technique, based on inspiration from the simulation of the biological brain [21]. ANN exhibited a good fit of the experimental data with high adjusted coefficient of determination (R^2_{adj} = 0.885), as well as high accuracy of the estimation (MSE=0.054). The difference between the experimental and estimated average color properties (residuals) vs the experimental are shown in Fig 3. It was observed that 72.22% of the results were between -0.11337 and 0.16324. Thus, corroborate the suitability of the proposed model to accurately predict the color properties of ulluco edible films.

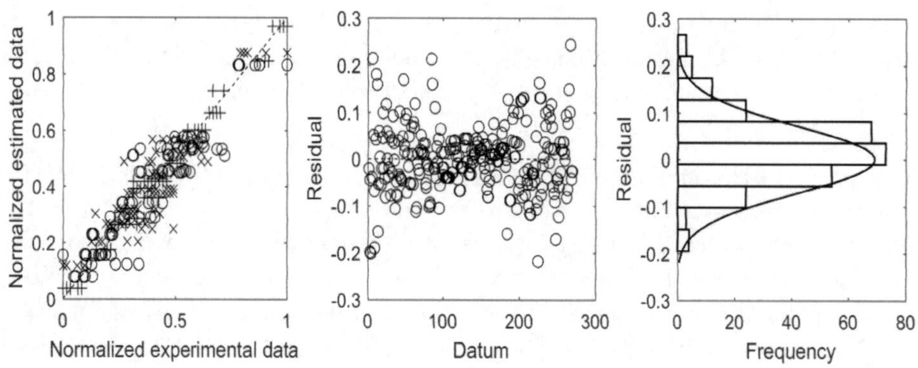

Figure 3. Residual analysis for color properties of ulluco edible films to ANN

A mathematical model for the feed-forward architecture was obtained (Equation 3).

$$\mathbf{Y} = \mathbf{W}_{ho} \times \left\{ \frac{1}{1 + \exp\left[-\left(\mathbf{W}_{ih} \times \mathbf{X} + \mathbf{b}_h\right)\right]} \right\} + \mathbf{b}_o \quad (3)$$

where $\mathbf{W}_{ih}$ is the weight matrix from the input to the hidden layer, $\mathbf{b}_h$ is the bias vector of the hidden layer, $\mathbf{W}_{ho}$ is the weight matrix from the hidden layer to the output layer, $\mathbf{b}_o$ is the bias vector of the output layer, **X** is the matrix of the normalized input variables and **Y** is the matrix of the normalized output variables.

4. Conclusions

Edible films based on ulluco starch were successfully prepared and the effect of starch concentration *SC*, glycerol concentration *GC* and temperature *T* on their color parameter (L*, a*, b*) was study using an artificial neural network ANN. Colorimetric analysis show high L* values, neutral a* values and slightly yellowness films (positive b* values). L* values were mainly influenced by *SC* and by the interactions between *SC* and *GC*. Nevertheless, a* and b* values were mainly influenced by *SC* and by the interactions between *SC* and *T*. ANN accurately predicts the color properties of ulluco edible films. Thus, ulluco edible films has suitable color characteristics for being applied as for food coating or edible application, corroborated by their high transparency and luminosity.

References

[1] Lima M et al 2017 Structural, Thermal, Physical, Mechanical, and Barrier Properties of Chitosan Films with the Addition of Xanthan Gum *J Food Sci* **82**(3) p 698-705.
[2] Baron R D et al 2017 Production and characterization of films based on blends of chitosan from blue crab (*Callinectes sapidus*) waste and pectin from Orange (*Citrus sinensis* Osbeck) peel *Int J Biol Macromolec* **98** p 676-683.
[3] Nawab A et al 2017 Mango kernel starch-gum composite films: Physical, mechanical and barrier properties *Int J Biol Macromolec* **98** p 869-876.
[4] Ali A et al 2017 Preparation and characterization of starch-based composite films reinforced by corn and wheat hulls *J Appl Polym Sci* **134**(32) 45159.
[5] Costa M J et al 2015 Use of wheat bran arabinoxylans in chitosan-based films: Effect on physicochemical properties *Ind Crops Prod* **66** p 305-311.
[6] Basiak E Lenart A and Debeaufort F 2017 Effect of starch type on the physico-chemical properties of edible films *Int J Biol Macromolec* **98** p 348-356.
[7] Cejudo-Bastante M J et al 2014 Potential use of new Colombian sources of betalains. Color stability of ulluco (*Ullucus tuberosus*) extracts under different pH and thermal conditions *Food Res Int* **64** p 465-471.
[8] Viehmannová I et al 2012 Induced polyploidization and its influence on yield, morphological, and qualitative characteristics of microtubers in Ullucus tuberosus *Plant Cell Tissue Organ Cult* **109**(1) p 83-90.
[9] Aguirre Forero S E et al 2012 Andean Tuber Production System in Boyacá, Colombia *Cuad Desarro Rural* **9**(69) p 257-273.
[10] Cruz G et al 2016 Physicochemical and rheological characterization of Andean tuber starches: Potato (*Solanum tuberosum ssp. Andigenum*), Oca (*Oxalis tuberosa* Molina) and Papalisa (*Ullucus tuberosus* Caldas) *Starch Starke* **68**(11-12) p 1084-1094.
[11] Valcárcel-Yamani B Rondán-Sanabria G G and Finardi-Filho F 2013 The physical, chemical and functional characterization of starches from Andean tubers: oca (*Oxalis tuberosa* Molina),

olluco (*Ullucus tuberosus* Caldas) and mashua (*Tropaeolum tuberosum* Ruiz & Pavón) *Braz J Pharm Sci* **49**(3) p 453-464.

[12] Tumwesigye S et al 2016 Novel Intact Bitter Cassava: Sustainable Development and Desirability Optimisation of Packaging Films *Food Bioprocess Tech* **9**(5) p 801-812.

[13] Ticknor J L 2013 A Bayesian regularized artificial neural network for stock market forecasting. *Expert Syst Appl* **40**(14) p 5501-5506.

[14] Villa-Vélez H A Váquiro H A and Telis-Romero J 2015 The effect of power-ultrasound on the pretreatment of acidified aqueous solutions of banana flower-stalk: structural, chemical and statistical analysis *Ind Crops Prod* **66** p 52-61.

[15] Pantoja D C et al 2016 Procesamiento de Arvejas (*Pisum sativum* L.). Parte 1: Modelado de la Cinética de Secado por Capa Delgada de Arveja, Variedades Obonuco Andina y Sureña *Inf Tecnol* **27**(1) p 69-80.

[16] Sepúlveda, F D Cisternas L A and Gálvez E D 2014 The use of global sensitivity analysis for improving processes: Applications to mineral processing *Comput. Chem. (Oxford)* **66** p 221-232.

[17] Maniglia B C et al 2017 Bioactive films based on babassu mesocarp flour and starch *Food Hydrocoll* **70** p 383-391.

[18] Damas M S P et al 2017 Edible films from mucilage of *Cereus hildmannianus* fruits: Development and characterization *J Appl Polym Sci* **134**(35) 45223.

[19] Saberi B et al 2016 Optimization of physical and optical properties of biodegradable edible films based on pea starch and guar gum *Ind Crops Prod* **86** p 342-352.

[20] Vargas C G et al 2017 Comparative study on the properties of films based on red rice (*Oryza glaberrima*) flour and starch *Food Hydrocoll* **65** p 96-106.

[21] Ameer K et al 2017 Optimization of microwave-assisted extraction of total extract, stevioside and rebaudioside-A from *Stevia rebaudiana* (Bertoni) leaves, using response surface methodology (RSM) and artificial neural network (ANN) modelling *Food Chem* **229** p 198-207.

IOP Conf. Series: Materials Science and Engineering **437** (2018) 012004 doi:10.1088/1757-899X/437/1/012004

Detection of palm oil bud rot employing artificial vision

D Montero[1] and C Rueda[1]

[1]Universidad Pontificia Bolivariana, Autopista Piedecuesta km 7, Bucaramanga, Colombia

claudia.rueda@upb.edu.co

Abstract. This paper describes the development of a system capable of capturing, processing, and analysing images of palm oil plantations in order to detect and identify bud rot disease. The process begins with the capturing of images using the DJI Phantom 4 UAV, which is first configured for a flight plan of the desired area and altitude. The resulting images are processed by photogrammetry software to create orthomosaics. The developed algorithm uses the grayscale of the generated orthomosaic and identifies the palms affected by bud rot. This is accomplished using the sliding-window method, by which smaller samples of the image are used and evaluated independently. Each sample is then extracted with a ULBP feature vector that numerically represents the texture of the image and is classified by means of a previously trained logistic regression model, allowing the recognition of positive cases of the disease. Possible positive cases are further distinguished using the non-maximum suppression algorithm. The system was tested with different images than the images used for training and for establishing the set point. As a result, the system showed a 92% precision and a 96% of sensitivity for bud rot disease detection. These results are satisfactory in terms of detecting bud rot disease using a low-cost system.

1. Introduction

Colombia is the largest producer of palm oil in Latin America and the fourth largest in the world. As such, palm oil is one of Colombia's principal and most valuable agricultural products. The principal threats to Colombia's palm oil industry are exposure to pests and disease. The most common disease is bud rot [1][2]. Bud rot can decrease the production of a given field from 8% to 29% [3], which results in millions of dollars lost annually. At this time, the oil palm industry in Colombia has a great opportunity to contribute to the development of the country. Therefore, it is important that significant investments be made in technologies that ensure plant health and increase yield, i.e., that improve overall crop quality and enhance sustainability of the industry [4]. Traditionally, disease detection, especially of bud rot, has been the responsibility of trained personnel who manually inspect all areas of a plantation for the diseased crop and then write this information on paper charts. To inspect an entire crop using this method can require months. This method is not only inefficient and unnecessarily time consuming, but it is also labor intensive, uneconomical, and prone to error. The process is so time-consuming that it is likely that the crop will change during the inspection process, rendering the results even less useful. Currently, some oil palm farmers use photos taken by small airplanes to show the health of the palm with respect to bud rot, observing the diameter of the crown [5], but this inspection method is prohibitively expensive due to the cost of flight time. For this reason, unmanned aerial vehicles (UAVs) have gained prominence in the monitoring of oil palm plantations. This new technology offers a wide variety of equipment designed to fly and capture aerial images over large areas at a reasonably low cost, with resolutions varying depending on the properties of the onboard cameras, altitude and flight autonomy [6].

Combining techniques of computer vision and machine learning with the images taken by UAVs have been implemented successfully to create a lower-cost, less labor-intensive way to identify pests in Begonia Semperflorens crops [7], as well as to count the palms in oil palm plantations [8]. This technique could be applied further in the detection and classification of diseases in the leaves of a variety of native plants found in Tamil Nadu [9].

The purpose of this paper is to fully outline the development of a system capable of detecting and identifying bud rot disease in a timely, efficient manner that will overall benefit production. Images are first captured by UAVs and then processed with photogrammetry software into a more usable format. These orthomosiacs can be analysed by a logistic regression model that can detect specific areas of bud rot disease among the palms.

2. Palm oil bud rot disease detection system

This system is based essentially on three stages. These stages are outlined as a block diagram in figure 1. The first stage is the acquisition of images using the DJI Phantom 4 UAV, followed by the creation of orthomosaics that will later be processed with an algorithm for detection of bud rot disease.

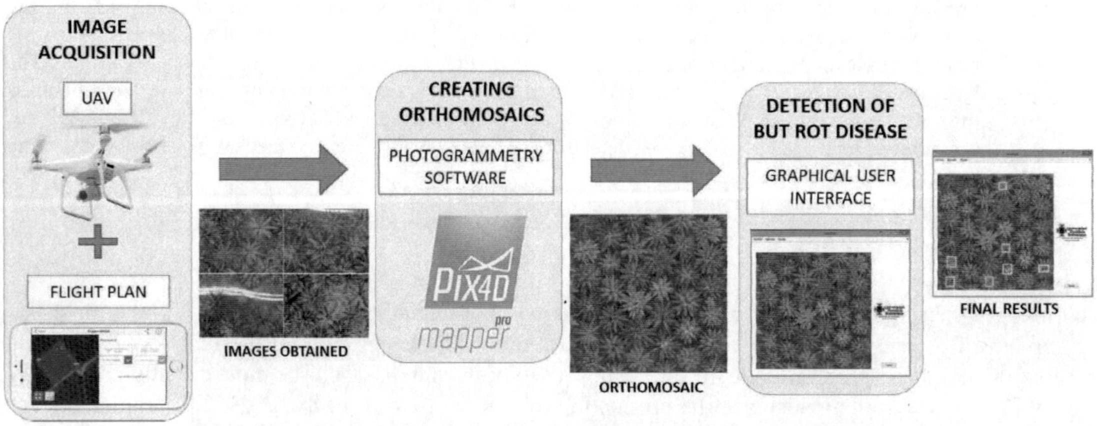

Figure 1. Block diagram of the general structure of the system

2.1. Image acquisition

Image acquisition is the first and one of the most important stages of the detection process as it significantly influences the subsequent stages. Without proper imaging, errors may result during the creation of the orthomosaic, thus preventing adequate detection of bud rot disease. The images are obtained with the Phantom 4 UAV, which uses a mobile application to create a flight plan at a programmed altitude, with a manually specified overlap. The flight plan controls the points where the images are captured. These images are acquired with the maximum frontal and lateral overlap, which is 90% at a height of 25 m. This overlap makes it possible to compensate for the perspective error caused by low-altitude flight. Figure 2 shows some of the images obtained in this stage. Low-altitude flight allows analysts to photograph the most relevant part of the palm at a quality sufficient for the detection of the bud rot disease.

2.2. *Creating orthomosaics*

The images resulting from the first stage are initially processed by the Pix4Dmapper photogrammetry software [10], via which all images are related and all geometric errors are corrected so that every point in the terrain is observable from a perpendicular perspective, as shown by the orthomosaic in figure 3. This process is performed automatically by finding thousands of common key points between the different images.

Figure 2. Photos obtained with the flight plan

Figure 3. Orthomosaic with constant orthogonal projection for every point

2.3. *Design of detection algorithm of but rot disease of the oil palm*

The development and evaluation of the algorithm used 15 orthomosaic, which were divided according to the recommendations of Ng [11] indicating that the database should be divided into three groups: 60% of the images for training, 20% for validation, and the remaining 20% for the evaluation of the model. The final images are completely different from those used in the process of creating the model.

2.3.1. Training of the logistic regression model. The system implements a supervised, machine-learning model [12][13][14] [15][16] for the detection of bud rot in diseased palms. This logistic regression model requires a training database classified into two categories: positive samples (diseased palms) and negative samples (healthy palms and backgrounds). A professional technician in oil palm Classification does this separation, focusing especially on differentiating between healthy and diseased palms. This process is performed in all the orthomosaics that were previously processed.

What resulted is a database that includes 450 positive samples and 2700 negative samples, as observed in Figure 4 using 60% training orthomosaics. These RGB samples are then converted to grayscale, followed by the application of a median filter in the spatial domain to reduce "noise". Every sample has a size of 750x750 pixels, thus ensuring that the center of the palm and its main branches can be observed. Each one is described using U-LBP where each pixel gets a value according to the texture patron that surrounds it. This descriptor has the quality of being invariant to the rotation. As a result, we modified the descriptor to detect the diseased palm in the center of each window. The new descriptor is made by the consecutive union of 9 U-LBP vectors, consisting of 59 patterns each. It is obtained by dividing the sample into 9 equal parts. As a result, we achieved a vector of 531 values per sample. Using these vectors, a matrix is created in which the rows correspond to each training image and the columns to the descriptor (figure 4). Finally, a preliminary logistic regression model is created with reference to the training matrix.

Figure 4. Examples of positive and negative samples from training database

2.3.2. Software for detection of the but rot disease in oil palms. The block diagram in figure 5 illustrates the process to detect bud rot disease in oil palm plantations.

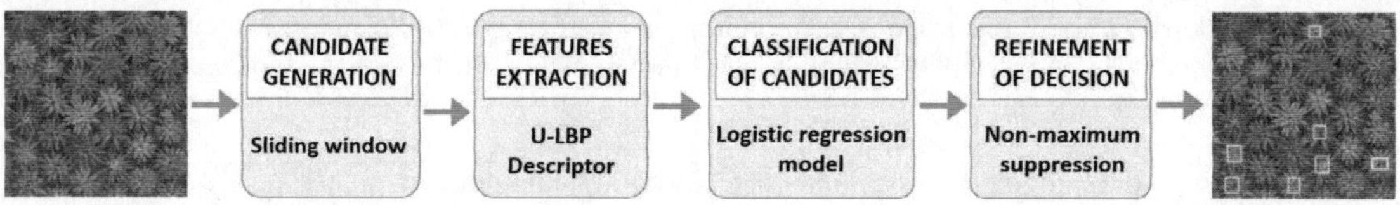

Figure 5. Block diagram of the software to detect bud rot disease.

First the candidates for bud rot must be generated. To do so, a sliding window is applied to the entire orthomosaic. These windows must have the same structure as the samples used in the training. For this reason, each candidate is shown in grayscale and is given a median filter for noise reduction. Also, each is sized at 750x750 pixels. The characteristics of each window are then extracted by means of the descriptor U-LBP. These characteristics are evaluated by the logistic regression model, which results in the generation of a "1," indicating a window containing a palm possibly diseased with bud rot, or a "0," indicating a healthy palm or mere background. Finally, it is necessary to refine the results because it is possible that the same diseased palm will be detected in more than one window.

The refinement is conducted by means of the non-maximum suppression technique.

2.4. Evaluation

2.4.1. Evaluation metrics. In the evaluation of the metrics, three orthomosaics corresponding to 20% of the total are used, as they are different to those used in the training of the model. This is done in order to adjust and refine the final logistic regression model. The first metric evaluated is the decision threshold of the logistic regression model, while the second is the overlap threshold of the non-maximum suppression technique. Sensitivity and precision were used as evaluation parameters. This process resulted in 0.5 for the decision threshold and 0.4 for the NMS threshold, respectively, as the best parameters to adjust the model.

2.4.2. Results. Finally, we used the remaining 20% of the orthomosaics for the evaluation of the system—those which were not used in the process of training or adjustment of the system—in order to evaluate and guarantee the performance for new plantation palms. The end result demonstrated a sensitivity of 96%, specificity of 97%, precision of 92% and a Kappa index of 92%, observed through the confusion matrix of figure 6.

		TRUE CONDITION		
	Prediction	Oil palm with bud rot disease	Healthy palm	**Total**
PREDICTED CONDITION	Oil palm with bud rot disease	23	2	25
	Healthy palm	1	76	77
	Total	24	78	102

Sensitivity	96%
Precision	92%
Kappa	92%

Figure 6. Confusion matrix of system evaluation

3. Discussion

The results yielded by this logistic regression model, using parameters of 0.5 for the decision threshold and 0.4 for the NMS threshold and a ULBP descriptor, confirm the model and descriptor originally promised by Mizerque [8].

This classification model obtained can be improved to make it more robust for the detection of bud rot disease in palm oil crops, by varying the parameters of the ULBP characteristic descriptor and comparing them by means of the ROC curves [17] and the method of cross-validation.

4. Conclusions

The high quality obtained in the orthomosaic generated by the Pix4Dmapper photogrammetry software allowed the developed classification algorithm to obtain the desired results with a sensitivity of 96% and a precision of 92%. Moreover, it was possible to prove that using UAVs is an economical method

IOP Conf. Series: Materials Science and Engineering **437** (2018) 012004 doi:10.1088/1757-899X/437/1/012004

to obtain good quality information from palm oil crops for the purpose of identifying bud rot, thus avoiding the labor-intensive effort of having workers perform periodic visual inspections by traversing the crop.

References

[1] Sanz J 2016 *Palmas.* Bud Rot: Comprehensive Confrontation of a Lethal Enemy, P. Palmivora **37** 109-110
[2] Chinchilla C 1998 *Palma.* **19** 242-6
[3] Croplife 2010 <http://www.croplifela.org/es/proteccion-de-cultivos/plaga-del-mes/pudricion-del-cogollo-pc.html>
[4] Silva J 2010 *Palmas.* The oil palm industry in the Americas **31** 245-246
[5] Khairunniza S 2014 *Int Conf. on Agriculture and Agricultural Science Procedia (Kuala)* **2** 156-164
[6] Rincon V, Molina A, Torres L and Herazo A 2015 *Palmas.* Outlook of UAV Technology in Oil Palm Plantations: Crop Monitoring through High-resolution Aerial Images **36** 25-41
[7] Amaya O, Caceres C and Ramos O 2014 *Agron* **64** 273-275
[8] Miserque J, Laverde R and Rueda C 2016 *IOP Conf. Series: Materials Science and Engineering.* Development of an aerial counting system in oil palm plantations **138** 2-6
[9] Arivazhagan S, Newlin R and Varthini S 2013 *Commission of Agricultural Engineering.* Detection of unhealthy region of plant leaves and classification of plant leaf diseases using texture features. **6** 211-217
[10] Diaz C 2013 Adquisición de imágenes de bajo costo aplicadas a la agricultura de precisión usando vehículos aéreos no tripulados (Quito: Universidad San Francisco)
[11] Ng P 2013 *Palmas.* **34** 259-79
[12] Neupane R, Sharma K and Thapa G 2002 *Agricult. Syst.* **72** 177-96
[13] Ohlmacher G and Davis J 2003 *Engin. Geology.* **69** 331–43
[14] Stephenson C, MacKenzie M, Edwards C and Travis J 2006 *Ecological Modelling* **193** 747- 58
[15] Mattila U and Nuutinen T 2007 *Silva Fennica.* **41** 29-43
[16] Mila A, Carriquiry A and Yang X 2005 **94** 102-110
[17] Rueda C, Miserque J and Laverde R 2016 *IEEE ANDESCON*

IOP Conf. Series: Materials Science and Engineering **437** (2018) 012005 doi:10.1088/1757-899X/437/1/012005

Drying of Yacon (*Smallanthus sonchifolius*) as a potential food product for international commercialization

S P Cuervo[1], A Benitez[2,3], S M Castellanos[4]

[1] Universidad Pontificia Bolivariana. Research Group GIDETECHMA. Bucaramanga, Colombia

[2] Universidad UDES. Research Group GEDETEC. Bucaramanga, Colombia

[3] Doctorate in Statistics, Institute of Applied Statistics and Information Technology. University of the Andes, Mérida, Venezuela

[4] Universidad Pontificia Bolivariana; Bucaramanga, Colombia

sandra.cuervo@upb.edu.co

Abstract. This paper describes the drying process conditions with the objective to minimise the negative effect over the final quality of the dry Yacon (*Smallanthus sonchifolius*) for international commercialization. Yacon is a tuberous root that grows throughout the Andean mountains in different South American countries. During the last years the interest for this product has been increasing and researchers have collected information that indicates the great potential of yacon. The product cultivation has been promoted over the past few years, due to the health benefits found after consumption. This agricultural product contains FOS (Fructo Oligo Sacarides) of low molecular weight, which is used as alternative sweeteners and, due to the small amount of calories is also attractive and good for overweight and diabetic people. Due to the outstanding importance of its health properties, this studied was carried out to achieve the best parameters of the drying process. For this paper, an experiment was performed considering two factors of the drying process: thickness of the layer (2cm, 4cm and 6cm) and temperature (60°C, 70°C and 80°C), with measurements of the physicochemical properties in fresh, before drying and after the drying process, with three replicates for each combination of levels.

1. Introduction

Yacon is a tuberous rhizome that growths in the Andean region spreading throughout the humid mountains of Peru, Colombia, Venezuela, Ecuador, Argentina and Bolivia [1]. It is a high-producing crop that yields an average of 30 tons per hectare and has a lifetime of 15-20 days in normal conditions. Yacon is most commonly consumed fresh, in juices, extracts or capsules, and in cooked dishes [2]. The yacon contains FOS (Fructo Oligo Sacarides) of low molecular weight, normally used as alternative sweeteners with small amounts of calories. It is attractive and is good for overweight and diabetic people since the human body cannot absorb the sugars it contains. The sugars are called inulin and oligofructose, and contribute to a healthy diet in line with the World Health Organization's recommendations regarding nutrition, physical activity and health [3].

Global circumstances and new directions in food consumption are seen as essential components in order to try to find opportunities of commercialization of fruits and vegetables in Latin American countries [4]. The demand of agricultural products by countries around the world has increased over the years and has shown great opportunities to agricultural products like yacon, which is part of all the potential products in developing countries such as Colombia [5]. A need for conservation of food in different countries has also open a new way of consumption and a demand for processed food [6]. The significant interest of people with purchasing power is focused on natural and processed superfoods, which can assure easy consumption for a healthy diet [7].

International markets register an increase on the demand of functional food, forcing countries around the world to increase and diversify the agricultural products [8]. The production and distribution of functional products is risky, expensive and sometimes difficult, due to the individual requirement of the global commerce [6]. A way to minimize the risk is by processing and extending the life time of the products by using different agro industrial transformation processes [9].

Drying is one of the oldest processes used in the conservation of foods with notable effects on the quality of the product [10]. Not all products have the same features in the drying process [11]. That is why it is so important both to carry out research to define the drying parameters of each product in order to minimize the negative effects on quality, and finding the drying conditions that guarantee a quality product competitive in the international market [12]. Others studies in the costumer perceptions have already gave this researchers a starting point in the objective of a product that have good acceptance in a global level of commercialization [13].

2. Methodology

2.1. Materials

Yacon roots (*Smallanthus sonchifolius*) were purchased in local markets in the city of Bucaramanga (Colombia), and were selected taking into account similarity in both color and firmness. The yacons were washed and peeled, and immediately rinsed with cold water to remove the remains.

After washing the yacon it was cut into slices of 2 mm, 4 mm and 6 mm, immersed immediately in a solution of citric acid at 1.25% for 2 minutes and then put over paper towels during one minute.

2.2. Experimental design.

To evaluate the effect on the quality characteristics of the factors, drying temperature (levels 60,70 and 80 °C and thickness (levels 2, 4 and 6 mm), a factorial design of 3^2 was constructed. The measurement of the effect of the 9 treatments was carried out with three replicates and, to homogenize the environmental factors on the drying process, the order of the resulting 27 replications was randomized (9x3). To guarantee the homogeneous samples of the product used in each replica, the initial characterization of the fresh product was measured and compared before statistical analysis of the treatments.

2.3. Moisture content:

The moisture content was determined according to norm AOAC 2013, using the oven method, that is, to carry a sample of 100 grams of (*Smallanthus sonchifolius*) at 103 °C +- 2 to constant weight and then using equation $H = \frac{W_i - W_f}{W_i} * 100$ where W_i is the weight of the solid + water (total kg of water plus dry solid) before the drying process, W_f is the weight of the sample after the drying process. The moisture content was determined before the drying process in triplicate, and to ascertain moisture content at the end of the drying process estimations were carried by in triplicate and by combinations of the levels of the factors.

2.4. Color:

For color measurement, an expandable polystyrene (EPS) made chamber was used, with an opening in the upper part to guarantee that the light input and the distance to the sample were always the same. The measurements were made with a Colorimeter Model CR400, Konica Minolta through the CIELAB system. The measurements were taken at the same time of the day to reduce variability. The colorimeter was calibrated with a white pattern. The values of the coordinates were taken; L*(dark - light), a* (green - red) b * (blue - yellow) and was calculated ΔE, by means of the following equation:

$$\Delta E = \sqrt{(\Delta a^*)^2 + (\Delta b^*)^2 + (\Delta L^*)^2} \quad (1)$$

The system used to measure the color takes into account three parameters: Lightness (L*), saturation (C*) and tone (h*). Subsequently, the fresh samples were compared with the samples submitted to the different drying treatments [14].

$$C^* = \sqrt{(a^*)^2 + (b^*)^2} \tag{2}$$

$$h^* = tan^{-1}\left(\frac{b^*}{a^*}\right) \tag{3}$$

2.5. Sugars content: sucrose, fructose, glucose and inverted sugar

10 milliliters of the juice of the pulp was extracted by compression of Xg of yacon pulp in a container. The types of sugars were measured by means of 4 refractometers: %Brix, %fructose, % glucose and % inverted sugar (Hann instruments 96801 / Hann instruments 96802 / Hann instruments 96803 / Hann instruments 96804 respectively). Tests percentages were done by triplicate.

2.6. Drying process

For the drying tests, a Binder FD 115 forced convection oven was used. For the drying loss curves, moisture loss control is carried out, weighing the samples throughout the entire process. The first hour the samples are weighed every 15 minutes, then every half hour for 2 hours and then every hour until constant weight is achieved.

During the process of drying, the color change and size of yacon slices and weight loss were recorded.

2.7. Volatile Organic Compounds (VOCs) from Yacon fresh and after a drying process Extraction by Headspace-Solid Phase Micro Extraction (HS-SPME).

The solids obtained from yacon slices fresh and after drying were homogenized and placed in a 10ml vial. The vial was sealed by an air-tight Teflon septum, an aluminum cap and incubated at 30ºC for 15 min in a water bath. A 1-cm SPME fiber (50/30um DVB/CAR/PDMS; Supelco, Bellefonte, PA, USA) was manually inserted into the headspace of the sample vial and exposed for 15 min. The volatile compounds were thermally desorbed in the GC injection port (splitless mode) for 5 min at 250ºC.

2.8. Gas Chromatography-Mass Spectrometry (GCMS) analysis

GCMS analyses of the SPME fiber were conducted using a Shimadzu GC5050 chromatograph coupled to mass spectrometer. After desorption of volatile compounds from SPME fiber, the analyses were separated made on a HP-5 column (30m x 0,25mm i.d., 0,25 um df) (Agilent Technologies, USA) and a DB-WAX column (30m x 0,25mm i.d., 0,32 um df) (J&W Sci) according with the following temperature program: started at 50ºC, then increased to 190ºC at a rate of 6ºC/min, and then to a 240ºC at a rate of 12ºC/min, with a final hold time of 2 min. Helium was used as a carrier gas at a flow rate of 1,5ml/min, injector and detector temperature, 250ºC. MS was operated in electronic impact (EI) mode 700eV, with mass range 40-350 atomic mass units, a source temperature 280ºC, and transfer line temperature 250ºC.

2.9. Volatile compound identification

Tentative identification was attempted by comparison of the retention index (RI) and mass spectra against NIST spectrum data base using NIST MS Search 2.0 (NIST, Gaitherburg, MD, USA). Retention indices (RI) were calculated using a homologous series of C6-C25 alkanes standard solution. Chemical authentic standards, when available, were analyzed under the same chromatographic conditions and their RI and mass spectra confirmed compounds identities.

2.10. Statistical analysis

The data collected on the different characteristics of interest were described by averages and medians (measures of central tendency), standard deviations (SD) and coefficients of variation (homogeneous measures CV% <10%) in the case of quantitative variables. To identify the shape of the distribution of the variables, asymmetry and kurtosis coefficients were used. Between treatments comparisons were made with the test of medians.

IOP Conf. Series: Materials Science and Engineering **437** (2018) 012005 doi:10.1088/1757-899X/437/1/012005

3. Results

Rhizomes were purchased in the last six months of study in different local markets, searching products with similar characteristics. See Figure 1, below.

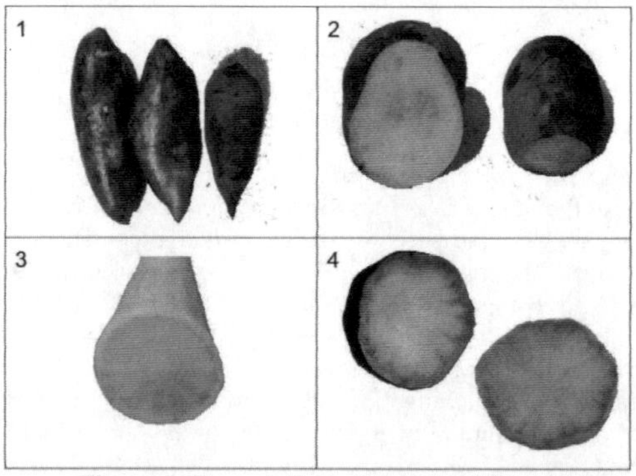

Figure 1. Presentation of the tuber Yacon. Notes: 1 tuber available in local markets. 2 Color and internal appearance of the Yacon. 3 Peeled Yacon ready for consumption or processing. 4 Yacon cuts for the drying process.

Lengths of Yacon varied between 6 and 31cm in the studied sample, with external shades of light brown, dark brown and purple. Internally the tubers presented diverse colors, as shown in figure 2.

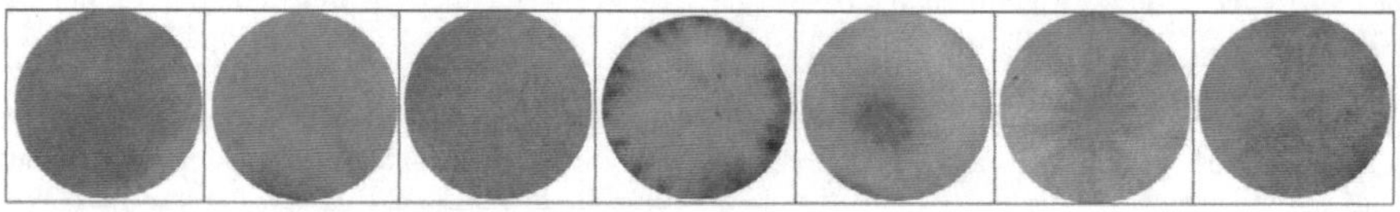

Figure 2. Shades of Yacon in fresco.

Yacon showed very homogeneous initial moisture values, with an average moisture content, and a dispersion percentage of only 6.4% with respect to their mean. The minimum value of moisture found was 73.94% and the highest value was 93%. Table 1. presents the percentage average values of glucose, inverted sugar, fructose and Brix sugar, with coefficients of variation higher than 20%, indicating a high dispersion of these measurements among the different analyzed roots. Despite the dispersion of the sugar measurements, the behavior of the variables is normal, according to the Shapiro-Wilk test (See Table 1).

Table 1. Description of the initial characteristics of the fresh yacon

Measurements	Moisture content	Glucose	Inverted sugar	Fructose	Brix
Promedio	84,72%	11,81%	12,34%	12,76%	13,29%
(CV%)	(7,07%)	(27,8%)	(26,17%)	(25,93%)	(24,13%)

To establish the initial homogeneity of the product conditions, the initial moisture content between the different treatments was compared, using the non-parametric median comparison test (see table 2).

Table 2. Moisture content of Yacon under the assignment process

Teatments	*%Moisture content (initial)*	*%Moisture content (final)*
60°C 2mm	84,4	1,66
60°C 4mm	86,3	1,70
60°C 6mm	86,3	1,79
70°C 2mm	85,5	1,35
70°C 4mm	75,9	1,58
70°C 6mm	80,7	1,44
80°C 2mm	93,0	0,82
80°C 4mm	83,7	1,09
80°C 6mm	80,1	1,35
$X^2_{(8)}$ *(p<0,05)*	*4,33 (0,836)*	*12,667 (0,027)*

The initial moisture content did not show significant differences according to the test of the medians comparisons ($X^2_{(8)}$ = 4.33, p = 0.836), while the medians of the moisture content in the dry Yacon offered significant differences($\chi^2_{(gl=8)} = 12{,}667;\ p = 0{,}027$). As observed in table 2 the moisture content of the dry Yacon at 60°C did not show differences between the diverse thicknesses compared ($\chi_{(gl=2)} = 0{,}9; p = 0{,}638$), likewise at 70°C; however, at 80°C the slices of 6mm presented higher moisture content than the slices of 2mm and 4mm ($\chi_{(gl=2)} = 6{,}3; p = 0{,}043$).

Table 3. Time of drying process under assignment conditions

Treatments	*Time (hours)*[a]
60°C 2mm	13,44
60°C 4mm	14,40
60°C 6mm	14,59
70°C 2mm	12,00
70°C 4mm	12,50
70°C 6mm	12,46
80°C 2mm	7,00
80°C 4mm	7,50
80°C 6mm	9,50
$X^2_{(8)}$ *(p<0,05)*	31,11 (<0,001)

Note. a= median of time.

Regarding drying time, significant differences were observed in the median time taken by the Yacon slices to find the equilibrium moisture, which is affirmed with a confidence level of 99% (see table 3). At 60°C half of the samples elapsed less than 14.4 hours while the other half had a longer drying time. In the case of the samples dried at 70°C, the drying time of 50% of the samples was less than 12.5 hours and, finally the samples dried at 80°C were 8 hour in the oven.

When comparing the observed drying time as a function of the temperature factor, it was found that the median drying time at 80°C is significantly lesser than the time needed to dry the product at 60°C 70°C ($\chi^2_{(gl=2)} = 11{,}22; p = 0{,}003$). No significant differences were observed between the required drying time at 60°C and 70°C.

The loss of moisture content of the Yacon presented greater dispersion in the slices dried at 60°C and the most representative variations were observed up to five hours after entering lthe convection oven; for drying at 70°C, the time in which Yacon shows the greatest moisture loss is four hours. Yacon slices dried at 80°C drop moisture with a higher proportion until three hours. Under the different drying temperatures used, the thickness of 2mm was always the one registering the highest moisture loss in the shortest time; the thicknesses of 4mm and 6mm under the temperature of 70°C showed well differentiated moisture loss, while the thicknesses of 4mm and 6mm at 80°C showed a very similar index of moisture loss (See figure 3).

The color of Yacon was recorded at the beginning and at the end, using coordinates L* a* b*. It was observed that the coordinates a* and b* showed greater variability (CV%) at the beginning. The lightness (L*) is presented as the most homogeneous characteristic recorded. The coordinates showed a symmetrical behavior with respect to the average (asymmetry less than 1 in absolute value) and a form of meso-acoustic Gaussian bell (kurtosis less than one in absolute value) as observed in table 4.

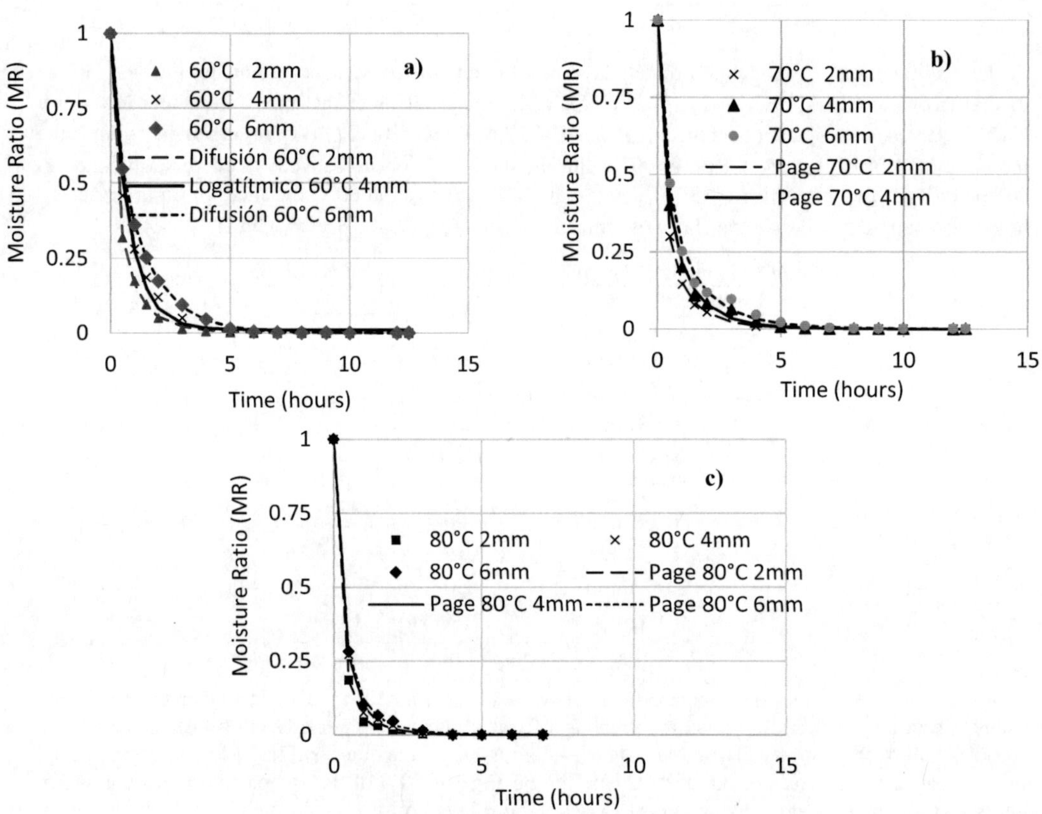

Figura.3. a. Curves of the drying process at 60°C. **b** Curves of the drying process at 70°C. **c** Curves of the drying process at 80°C

As seen in table 4, the values of the color coordinates L*, a*, b* of the fresh yacon and, after the drying process, it can be affirmed that there was a color change due to the drying process. Initially the fresh yacon is more luminous and greenish than the dry yacon.

Table 4. Yacon initial color representation using coordinates L* a* b*

	Coordinates	Median	±DS	CV%	Asymmetry	Kurtosis
Initial	L*	61,52	4,488	7,29%	-0,115	-0,724
	a*	2,277	1,852	81,3%	-0,769	-0,301
	b*	20,083	6,015	29,9%	0,348	1,278
Final	L*	45,898	6,704	14,6%	-0,148	-0,072
	a*	55,134	8,245	15,0%	-0,061	-0,888
	b*	0,561	0,155	27,6%	-0,167	-0,44

Notes. L* ligthness. a* coordinates red/green (+a indicates red, -a indicates green). b* coordinates yellow/blue (+b indicates yellow, -b indicates blue). sd; standard dev. CV%; percent coefficient variation.

The temperature of the drying air has a significant effect on the color characteristic of the Yacon. The parameter that had the highest sensitivity was L* (lightness). The value varies on average between 61 for fresh and 46.5 for dry, having a decrease of 25% represented in a darkening, as can be seen in Figure 4. With respect to the parameters a* and b* it is observed that the green color was accentuated, the values varied from 3 to 8. For the parameter b, the values varied from 21 and 28, accentuating the yellow color.

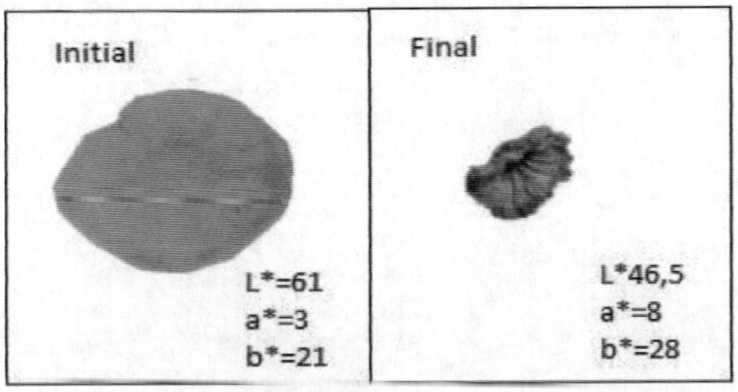

Figure 4. Color change in the yacon wheels of 6mm, in the process of drying the yacon at 80 ° C

To establish the change in the coloration of the product with the drying process, the median comparison test was used, finding no significant differences in the color change in the L* and b* axes (see table 5). In the coordinate a*, the color presented variations marginally significant, which implies that the magnitudes of the variations are slightly different between the treatments used ($p \approx 0.05$).

When analyzing the color changes that occurred due to the drying process, it was observed that the greatest color change was presented for the combination of 60°C and 4mm with a value $\Delta E = 39$ followed by the combination 60°C and 6mm with a value of $\Delta E = 34$, the lower temperature and greater thickness of the yacon slice, which leads to a longer drying time favoring browning; this is comparable with that reported by Cuervo-Andrade [14]. The smallest color change was presented for the combination of temperature parameters of 70° C and 2mm thickness.

The negative values of ΔH^* for all combinations of parameters indicate that the drying process increased the yellow color in the samples analyzed, the values between -1.26 and -1.45. In the case of saturation, it is observed that for all combinations of parameters the effect of the drying process was a decrease in saturation, which is reflected in the opaque color of the yacon slices as can be seen in Figure 4.

IOP Conf. Series: Materials Science and Engineering **437** (2018) 012005 doi:10.1088/1757-899X/437/1/012005

Table 5. Differences of medians in the initial and final color, coordinates L* a* b*

Treatments	ΔL^*	Δa^*	Δb^*	ΔE*	ΔC*	ΔH*
60°C 2mm	2,00	-4,33	-8,33	24,68	-42,55	-1,26
60°C 4mm	37,33	-9,00	0,33	39,05	-40,13	-1,26
60°C 6mm	30,07	-11,00	-9,33	34,11	-37,36	-1,38
70°C 2mm	8,67	-1,67	5,33	19,05	-40,00	-1,43
70°C 4mm	10,67	-7,00	-11,33	22,06	-29,40	-1,43
70°C 6mm	21,33	-10,33	-7,17	25,74	-31,93	0,06
80°C 2mm	15,67	-4,17	-13,33	23,96	-29,82	-1,45
80°C 4mm	16,67	-7,33	-7,67	19,12	-28,12	-1,41
80°C 6mm	8,67	-2,67	-9,67	25,11	-24,61	-1,45
$\chi^2_{(df=8)}(p)$	10,86 (0,210)	15,1 (0,058)	8,33 (0,400)	8,267 (0,408)	12,13 (0,105)	13,3 (0,083)

Nota. Δ = difference. χ^2= statistic Ji-square. df=degrees of freedom. p=p-hypothesis value. L*=ligthness. a*= coordinates red/green (+a indicates red, -a indicates green). b*= coordinates yellow/blue (+b indicates yellow, -b indicates blue).

When comparing the median value of ΔE* between the different treatments, it was observed that there is no significant difference in color variation, since within each treatment the median is similar to the overall median of the set of measurements, regardless of the treatments employed, affirmed with a confidence of 95%. A similar behavior was observed in ΔC* and ΔH*, color parameters that did not show important variations between the drying treatments used.

Table 6. Comparison of VOCs from Yacon fresh and after drying process at 60ºC at different thickness

No.	Compound	RI (HP-5)	RI (DB-wax)	%Area Fresh	4mm	6mm
1	acetic acid	600	1450	11,0	ND	ND
2	2,2,dimethyl-1-propanol	640	<1000	0,2	2,5	1,7
3	1-butanol	675	1145	13,1	3,2	1,9
4	2-methyl-1-butanol	720	1210	ND	0,3	1,7
5	4-hidroxy-4-methyl-2-pentanone	820	1352	0,1	15,2	17,9
6	heptanal	903	1174	ND	0,2	0,6
7	α-pinene	939	1032	0,5	ND	ND
8	camphene	953	1075	0,5	0,6	0,7
9	benzaldehyde	960	1495	1,2	11,6	17,3
10	β-pinene	981	1116	ND	4,3	3,9
11	butyl butanoate	985	1223	0,3	18,5	19,7
12	β-myrcene	992	1145	2,4	ND	ND
13	1-octanal	1006	1282	2,2	27,6	31,5
14	E-β-ocimene	1030	1230	49,7	ND	ND
15	dibutyl carbonate	1113	<1000	ND	33,3	28,5
16	2-metoxy-2-methylpropane	1130	<1000	ND	0.6	0,9
17	decanal	1208	1484	0,2	ND	ND
18	isoamyl decanoate	1615	1863	0,3	ND	ND

Note. ND Not detected

Volatile compounds of yacon pulp fresh and after hot air drying with three different sample masses and two different temperatures were extracted by Headspace-Solid Phase Micro Extraction (HS-SPME). Table 1 show the volatile compounds, retention indices on two orthogonal columns (HP-5 and DB-WAX) and concentration (Area % normalized). A total of 18 volatile compounds from yacon pulp fresh and dried powders were identified.

IOP Conf. Series: Materials Science and Engineering **437** (2018) 012005 doi:10.1088/1757-899X/437/1/012005

The major volatile compounds in fresh yacon were acetic acid, 1-butanol and E-β-ocimene. When the drying temperature increases, the amount of this major compounds decreases; and the minors almost disappear. After the drying process at 80ºC was not possible to find VOCs. So we compare only the influence of the thickness when the material is dried at 60ºC. In this way, a major thickness is better to avoid the loss of VOCs. In materials with 2mm was not possible to find VOCs.

4. Conclusions

The initialmoisture content of the Yacon did not present a significant difference which guarantees the use of homogeneous samples. As for the final moisture content, it can be said that increasing the temperature decreases the final moisture content of the product and, on the other hand, if the thickness increases, the moisture content of the product also rises.

Drying treatments do not significantly affect the final Yacon moisture index, and the color finally obtained for each of the treatments did not present significant differences due to the control of previous enzymatic browning and also because a loss of the volatile compounds was observed by drying effects. As for the conservation of the volatile compounds the Yacon dried at 60°C and 6mm slices was the best treatment.

The parameter that obtained the highest sensitivity due to the effect of the drying process was ligthness, with a variation of 25% with respect to the color of fresh yacon.

The combination of parameters with the less changes on the color was obtained for temperatures of 70°C and thickness of 2mm; but with these same parameters all the volatile compounds were lost.

References

[1]. Ministerio de Agricultura y Desarrollo Rural C. Anuario Estadístico del sector Agricopecuario Colombia 2013. 2014.

[2]. Choque Delgado GT, da Silva Cunha Tamashiro WM, Maróstica Junior MR, Pastore GM. Yacon (*Smallanthus sonchifolius*): A Functional Food. Plant Foods Hum Nutr. 2013;68(3):222–8.

[3]. Gusso AP, Mattanna P, Richards N. Yacon: Health benefits and technological applications | Yacon: Benefícios à saúde e aplicações tecnológicas. Cienc Rural. 2015;45(5-10).

[4]. D. M. U. P. Rocha, P. V. D. M. Ribeiro, A. P. S. Caldas, B. P. da Silva, A. da Silva, A. P. de Almeida, N. B. M. da Silva, A. M. Machado, and R. D. C. G. Alfenas, "Acute consumption of yacon shake did not affect glycemic response in euglycemic, normal weight, healthy adults," J. Funct. Foods, vol. 44, no. February, pp. 58–64, 2018.

[5]. Annunziata A, Vecchio R. Functional foods development in the European market: A consumer perspective. J Funct Foods. Elsevier Ltd; 2011;3(3):223–8. Available from: http://dx.doi.org/10.1016/j.jff.2011.03.011

[6]. Jongwanich J. Impact of food safety standards on processed food exports from developing countries. ADB Econ Work Pap Ser. 2009;154:1–28.

[7]. Grunert KG. Food quality and safety: Consumer perception and demand. Eur Rev Agric Econ. 2005;32(3):369–91.

[8]. Alexandratos N, Bruinsma J. World Agriculture Towards 2030 / 2050 The 2012 Revision Proof Copy. 2012;(12):146.

[9]. Timmer CP. Agriculture and economic development revisited. Agricultural Systems. 1992.

[10]. Wu Z, Li W, Zhao L, Shi J, Liu Q. Drying characteristics and product quality of Lycium barbarum under stages-varying temperatures drying process. Nongye Gongcheng Xuebao/Transactions Chinese Soc Agric Eng. 2015;31(11-15).

[11]. Ali MA, Yusof YA, Chin NL, Ibrahim MN. Effect of different drying treatments on colour quality and ascorbic acid concentration of guava fruit. Int Food Res J. 2016;23.

[12]. Sablani SS. Drying of fruits and vegetables: Retention of nutritional/functional quality. Dry Technol. 2006;24(2).

[13]. Castellanos R SM, Cuervo-Andrade SP, Benitez A. Customer perception of dry yacon as potential funtional food for international commercialization. En: 2 CONGRESO INTERNACIONAL EN ADMINISTRACIÓN DE NEGOCIOS INTERNACIONALES – CIANI 2017. (27-29 Septiembre de 2017:Bucaramanga, Colombia) Memorias. Bucaramanga: Universidad Pontificia Bolivariana, 2017. 675 p.

[14]. Cuervo-Andrade, S.P.: *Quality oriented drying of Lemon Balm (Melissa officinalis.)*, Universität at Kassel - Fachbereich Okologische Agrarwissenschaften - FachgebietAgrartechnik, Dissertation, 2011

IOP Conf. Series: Materials Science and Engineering **437** (2018) 012006 doi:10.1088/1757-899X/437/1/012006

Electrospinning synthesis of Li-Fe-P/PAN based micro-nanofibers as precursors for $LiFePO_4$ cathode material in Li-ion fiber battery applications

M A Cortes[1,2], N Castano[1,2], E Garcia[1,2] and H V Martinez[1,2]

[1] Mechanical Engineering Department. Grupo de investigación sobre Nuevos Materiales (GINUMA), Universidad Pontificia Bolivariana, Medellín, Colombia

[2] Mechanical Engineering Department. Grupo de investigación Energía y Termodinámica (GET), Universidad Pontificia Bolivariana, Medellín, Colombia

[3] Nanotechnology Engineering Department, Universidad Pontificia Bolivariana, Medellín, Colombia

mariaalexandra.cortes@upb.edu.co

Abstract. The increase popularity of wearable devices and smart textiles are being held back partly by better energy storage systems, which are based on electrochemical phenomena. Research in novel materials with optimal morphologies for better electrochemical performance is one the paths being followed. Micro/nano fibers are interesting morphologies as they present high surface areas, less agglomeration and flexibility, compared to micro/nano particles. In this article, precursor fibers for $LiFePO_4$ cathode material, defined as lithium-iron-phosphorous/polyacrylonitrile (*Li-Fe-P/PAN*) precursor nanofibers, have been synthesized by the electrospinning method. The influence of electrospinning parameters such as voltage (V), flow rate (F) and distance between the collector and the tip of the syringe needle (D), on the morphology of the fibers have been studied mainly by scanning electron microscopy (SEM). Optimal morphological features such as less formation of beads, small fiber diameter, less polydispersity and high fiber alignment, were observed at V=14 kV, F=0.7 mL/h and D=14 cm. The occurrence of nanofibers on the precursor *Li-Fe-P/PAN* material was observed through SEM, while the energy dispersive x-ray spectroscopy (EDS) confirms the presence of the fundamental elements in the precursor nanofibers.

1. Introduction

Flexible electronics and wearable devices have gained enormous popularity with the emergence of products such as the Google Glasses, Apple Watch and Samsung Bracelet. Perhaps the main aspect holding back this revolution is the need of better energy storage systems. Li-ion batteries are the best energy storage technology currently available, principally because of their size reduction and the energy they can store, as compared with other technologies [1]. Despite its advantages, the current way in which commercial Li-ion batteries are produced, cannot effectively satisfy the working conditions that wearable products require, namely flexibility (bending, folding and stretching [2,3] miniaturization and adaptability. Besides within Li-ion battery components the cost of processing cathodes are one the major contributing factors [4], the safety and capacity in Li-ion batteries are largely limited by the cathode and research around the globe has mostly concentrated on the development of nanostructured cathode materials to improve their intercalation features [5]. $LiFePO_4$ (more known as LFP) cathodes are one of the most promising materials for Li-ion batteries due to its excellent electrochemical properties, good energy efficiency, but mainly because of its safety compared to other materials for electrodes [6]. This material stands out for its small change in mechanical properties during lithiation/delithiation, low cost, low toxicity and low environmental impact, which can be an option to meet the previously mentioned conditions for wearable devices [7].

'Wearable batteries' is the term used for batteries that can achieve the energy demand for the so- called smart textiles, in particular with applications that combine textiles and electronics (e-textiles) [8]. Ideally, this systems should be directly in the form of a fiber so that it can be integrated naturally during the weaving process [8]. However, the research in this strategic topic is still limited.

Initial attempts to investigate fiber storage focused on using carbon materials (graphene and carbon nanotubes –CNTs-) to develop light, weight and flexible supercapacitors [9–19]. However, compared with supercapacitors, Li-ion batteries have much higher energy density per weight unit. In 2011, Cu's group in Stanford integrated Li-ion battery electrode materials into a 3D porous textile conductor and obtained capacities of 150 mAh/g with an 88.5% capacity retention after 30 cycles [20]. Since then, several groups have contributed to this field and two thorough complete reviews on wearable fiber-shaped batteries have been published recently [21,22].

Among the morphologies reported previously for light and flexible electrodes, micro/nano fibers offer such possibilities and present advantages such as less agglomeration, high surface area, less diffusion distance for both Li^+ and e^-, among others [23–25]. Electrospinning is a novel and versatile technique, in which a precursor solution is subjected to a high voltage, making elongated jets in the form of micro/nano fibers [26–28]. The formation of fibers is determined by many operating parameters such as applied voltage, solution concentration, viscosity, surface tension, conductivity, and flow rate. The structure of the resulting fiber can be tuned by carefully modifying these parameters.

This paper describes the production of Li-Fe-P/PAN fibers (from here on "precursor fibers") by electrospinning, that will later be used for the formation of LFP cathode fibers. A detailed study will be shown on how the variation of the equipment parameters (voltage, flow rate and distance from nozzle to substrate), affects the formation of fibers with different morphological features. This is a preliminary step on the synthesis of full LFP cathode fibers, thus after calcination of the Li-Fe-P/PAN precursor fibers here obtained, a LFP stoichiometry with atoms arranged in an olivine structure is expected.

2. Materials and methods

2.1. Precursor solutions

Liquid precursor solution for the electrospinning process is a relevant stage, as its properties, together with the electrospinning parameters, define the final fiber morphology. The viscosity, surface tension and the electric field will determine the stretching of the solution. These properties are defined by a correct choice of the polymer/solvent set [29].

The liquid precursor solution was prepared using stoichiometric amounts of lithium acetate, $C_2H_3LiO_2$, iron (III) nitrate nonahydrate, $(FeN_3O_9)\ 9H_2O$, and phosphoric acid, H_3O_4P, as suggested by Toprakci, et al. [30]. These salts were mixed with polyacrylonitrile, $(C_3H_3N)_n$ (PAN) and disolved in a miscible solvent composed of dimethylformamide, $(CH_3)_2NC(O)H$ (DMF) and acetone, $(CH_3)_2CO$. The Li:Fe:P molar ratio in the solution was 1:1:1. The concentration in weight for each component was 4% for the PAN, 8% for the Li-Fe-P precursors, 70.4% for the DMF, and 17.6% for the acetone. All chemicals used were purchased from Sigma Aldrich.

2.2. *Electrospinning set up*

A scheme of the system used to produce the Li-Fe-P / PAN precursor nanofibers is shown in Figure 1. Here, the prepared solution, which was loaded into a syringe pump, was delivered into a reaction zone in a controlled way. When a high voltage is applied to the solution, the forming spherical droplet becomes deformed into the denoted Taylor Cone [31], due to the repulsion force between the surface charges. When the system reaches a critical voltage, a jet is formed at the apex of this cone. Once optimal conditions involving the solution and process parameters are met (i.e. when the repulsive electrostatic forces overcome the surface tension), this jet is further stretched as a fine fiber which finally reach the substrate holder [27], [29], [32].

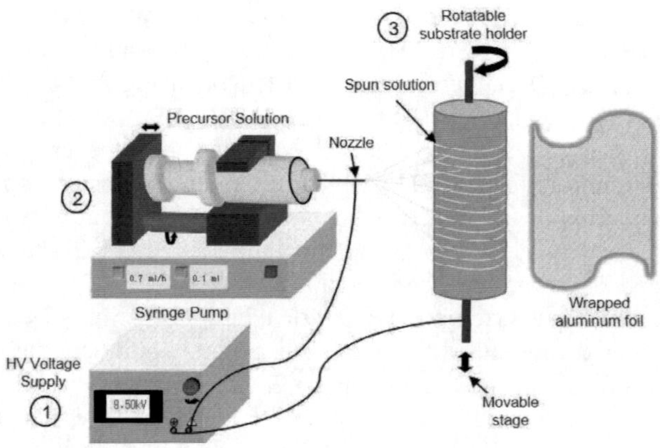

Figure 1. Schematic drawing of the electrospinning setup used to produce the Li-Fe-P / PAN precursor nanofibers. Note that the aluminum foil is rolled around the cylinder, the high voltage terminal is attached to the nozzle, while the ground is attached to the drum collector

2.3. *Morphological and compositional characterization*

Using the precursor solution described previously, a morphological study was then performed by varying only the parameters related to the electrospinning process. i.e. voltage (*V*), flux (*F*), and the distance between the tip of the nozzle and the substrate holder (*D*). Three levels were considered for each parameter. Table 1 shows the three levels for each electrospinning parameter. One parameter was varied while the other two were fixed. The fixed values were V=14 kV, D=10 cm and F=0.6 mL/h. The precursor fibers were collected on an aluminum foil as substrate. The fiber morphologies for each sample were investigated using a scanning electron microscopy (SEM: JEOL JSM-6010LA) operated at 5 kV. Compositional elemental mapping was investigated using an energy dispersive x-ray spectroscopy (EDS) detector coupled to the SEM (JCM-6000PLUS), operated at 15 kV. For all experiments, parameters such as 19G nozzle (outer diameter=1.07 mm / internal diameter = 0.69 mm), 19 rpm-clockwise drum rotation, temperature between 24-27°C and humidity between 55-65%, were fixed.

Table 1. Levels for each electrospinning parameter

V [kV]	*F* [mL/h]	*D* [cm]
14	0.5	10
15	0.6	11
16	0.7	12

3. Results and discussion

3.1. Changing voltage

Figure 3 shows the resulting fibers, using three different voltages in the electrospinning process: 14 kV (3a), 15 kV (3b) and 16 kV (3c). The fixed parameters in this case were the distance from the tip of the needle to the collector and the flow rate, with a corresponding value of 10 cm and 0.6 mL/h, respectively.

Figure 2. SEM micrographs showing the influence of voltage on the fiber morphology: (a) 14 kV, (b) 15 kV, and (c) 16 kV (F=0.6 mL/h; D=10 cm)

Fibers obtained at 14 kV have a clear alignment, in contrast with the fibers obtained at 15 and 16 kV, which present a curl-like shape. The curly effect obtained for these two last samples is due to the instability of the jet (whipping effect [33]), at higher electric fields (i.e. higher voltages) compared to the 14 kV sample [34]. Fibers produced at 16 kV confirm that alignment is lost, meaning the jet movement is higher for these conditions.

Regarding fiber diameter, the 14 kV sample have smaller fibers than those at 15 and 16 kV. Mats (the non-woven film of fibers collected on the counter electrode) produced at 14 kV present similar fiber diameters, with an almost monodisperse feature. In contrast, fibers produced at higher voltages present polydispersity in terms of diameter size.

The increase in diameter among samples might occur as the 15 and 16kV voltages are above the critical voltage. This may cause an incorrect Taylor cone development and a jet instability, which breaks the fibers during their pathway toward the collector, resulting in the formation of small curl-like fibers [28,35].

3.2. Changing distance

Figures 2a, b and c, show the influence in fiber morphology with distance values between the tip of the nozzle and the substrate of 10cm, 11cm and 12cm, respectively. The fixed parameters in this case were the voltage and the flow rate, with a corresponding value of 14 kV and 0.5 mL/h, respectively.

IOP Conf. Series: Materials Science and Engineering **437** (2018) 012006 doi:10.1088/1757-899X/437/1/012006

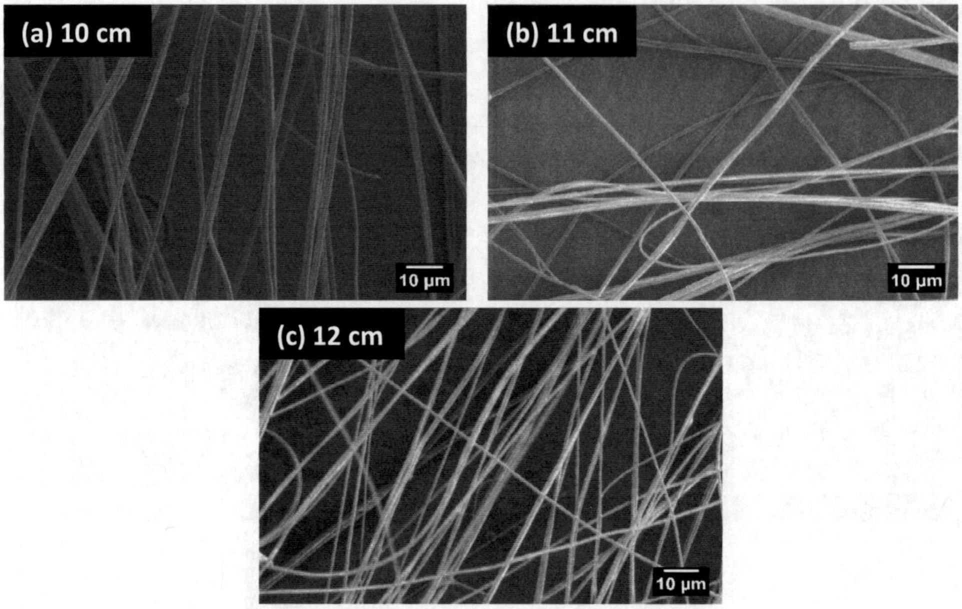

Figure 3. SEM micrographs showing the influence of distance from nozzle to substrate on the fiber morphology: (a) 10 cm, (b) 11 cm, and (c) 12 cm (V=14 kV; F= 0.6 mL/h)

In all three cases, there is a preferential alignment of the fibers, most likely in the direction of the rotation of the cylinder. There are no visible differences regarding the apparent diameter of the fibers. However, is difficult to determine the diameters of these fibers in the samples done at 10 and 11 cm, as they present more fiber agglomerations than the fibers observed in the sample done with a 12 cm distance. For this reason, the apparent diameter of fibers in 12 cm sample seems to be smaller and more monodisperse than those at 10 and 11 cm distances.

The strength of the electric field (E) depends on both the applied voltage and the distance ($E=\Delta V/d$). The latter also plays an important role on the time of flight of the jet [29], [36,37]. A shorter distance causes the time of flight to be reduced, therefore reduces the time for solvent to evaporate. Thus, fiber diameter is smaller for longer distances and tend to become larger for shorter distances. In addition, at shorter distances, the electric field strength is high, which causes the fiber to accelerate, resulting in a lower evaporation of the solvent [29]. When fibers with a large amount of solvent reach the substrate, the solvent tends to agglomerate the fibers.

In all cases presented previously, even though the solvents created fiber agglomeration, there are no observed droplets, which means that the DMF and the acetone have enough time to partially volatilize while the solution gets from the tip of the needle to the substrate.

3.3. Changing flowrate

The influence in fiber morphology with the variation of the flow rate is shown in Figure 4. Figure 4a corresponds to the sample obtained with 0.5 mL/h, while those of 0.6 mL/h and 0.7 mL/h correspond to Figures 4b and c, respectively. The fixed parameters in this case were the voltage and the distance from the tip of the needle to the collector, with a corresponding value of 14 kV and 14 cm, respectively.

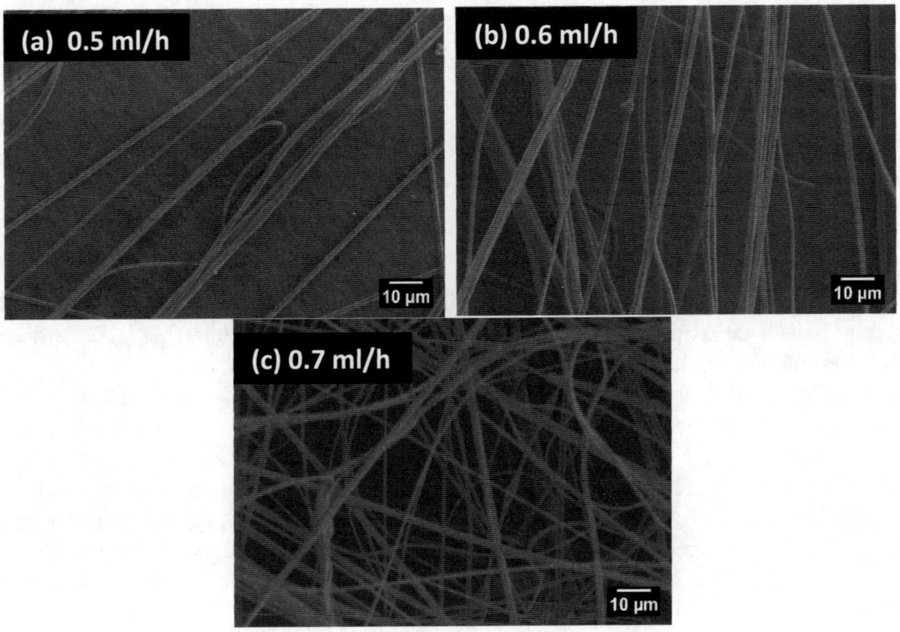

Figure 4. SEM micrographs showing the influence of flowrate on the fiber morphology: (a) 0.5 mL/h, (b) 0.6 mL/h, and (c) 0.7 mL/h (V=14 kV; D=10 cm)

For the fibers obtained with a flow rate of 0.6 mL/h and 0.7 mL/h, there are no significant changes in the apparent diameter. Conversely, the sample with F=0.5 mL/h apparently present a smaller fiber diameter. According to the literature the flow rate must exceed a critical value in which enough solution is fed to the tip of the nozzle before all the volume is drawn away by the strong electrostatic forces, causing a discontinuous or pulsating jet [38]. On the other hand, exceeding a certain flow rate may cause the formation of beads as there is not enough time for the extra solvent in the stretched fiber to evaporate [39]. As shown in figures 4(a-c), there is no beads, therefore it can be concluded that this upper critical flow rate was not achieved. Also, during the electrospinning process, for all the values investigated, there was no jet discontinuity, which means the flow rate was always above the lower critical flow rate. Moreover, it can be observed how an increase in flow rate leads to larger fiber diameters, which can be attributed to less evaporation of the solvent during flight from needle to collector [39]. Likewise, agglomeration of fibers becomes obvious when the solvent has not enough time to evaporate completely. Also, there is an apparent monodisperse distribution of the fiber diameters for the samples done with a flow rate of 0.5 mL/h and 0.6 mL/h, while that of 0.7 mL/h has presents different fiber diameters. However, fibers from the highest flow rate are less agglomerated than those fibers obtained with lower flow rates. Finally, the amount of fibers increases by increasing the flow rate, as the spread volume is higher.

3.4. Energy dispersive spectroscopy (EDS) analysis

EDS analysis was done for the Li-Fe-P/PAN precursor nanofibers. Table 2 shows the mass percentage for the elements present in the precursor nanofibers. It reveals the presence of carbon, oxygen, iron and phosphorus elements, which are the main elements to obtain the LFP active material. It is expected that after calcination of the Li-Fe-P/PAN precursor fibers here obtained, a LFP 1:1:1 stoichiometry with atoms arranged in an olivine structure is expected.

Table 2. Mass percentage of the elements present in the Li-Fe-P/PAN precursor nanofibers

Element	Mass Percentage	Atomic Percentage
Carbon	39,7254576	53,0606366
Oxygen	39,6145313	39,7234625
Phosphorus	5,56017748	2,88059916
Iron	15,0998336	4,33530174

4. Conclusions

Micro-nanofibers based on lithium, iron and phosphorous precursor salts, together with PAN polymer, have been successfully produced. This opens the door to obtain full LFP cathode fibers which will be employed on fiber battery applications.

The morphology and texture can be influenced by adjusting flow rate, voltage and distance from nozzle to sample of the electrospinning technique. According to the results discussed in this paper, to obtain monodisperse, aligned and straight Li-Fe-P/PAN precursor nanofibers, it is ideal to use a distance between the collector and the tip of the syringe of 12 cm, an electrospinning voltage of 14 kV, and a flow rate of 0.5 mL/h. It is important to clarify that these electrospinning parameters are recommended for the specific precursors used here, since properties of the solution such as concentration, viscosity, surface tension and conductivity will vary and hence, the optimal electrospinning parameters will also change.

It is relevant to mention that this is a preliminary step on the synthesis of full LFP cathode fibers. After calcination of the Li-Fe-P/PAN precursor fibers here obtained, a LFP stoichiometry with atoms arranged in an olivine structure is expected. Finally, the applicability of these fibers as cathode material will ultimately depend on their electrochemical performance.

This is an ongoing project and further work is needed.

Acknowledgements

The work carried out in this project has been financed by Universidad Pontificia Boliviariana and Patrimonio Autónomo Fondo Nacional de Financiamiento para las Ciencias, la Tecnología y la Innovación, Francisco José de Caldas. Dr. Erik Kelder is greatly acknowledged for granting permission to use the SEM facilities at ChemE - TU Delft.

References

[1] Manthiram A 2011 Materials challenges and opportunities of lithium ion batteries J. Phys. Chem. Lett. 2 176–84

[2] Liu B, Zhang J G and Shen G 2016 Pursuing two-dimensional nanomaterials for flexible lithium-ion batteries Nano Today 11 82–97

[3] Zhou G, Li F and Cheng H M 2014 Progress in flexible lithium batteries and future prospects Energy Environ. Sci. 7 1307–38

[4] Nitta N, Wu F, Lee J T and Yushin G 2015 Li-ion battery materials : present and future Biochem. Pharmacol. 18 252–64

[5] Cho S-J, Uddin M-J and Alaboina P 2017 Review of Nanotechnology for Cathode Materials in Batteries Emerging Nanotechnologies in Rechargeable Energy Storage Systems pp 83–129

[6] Jiajun W and Sun X 2012 Understanding and recent development of carbon coating on LiFePO 4 cathode materials for lithium-ion batteries Energy Environ. Sci. 34 1201

[7] Padhi A K, Nanjundaswamy K S and Goodenough J B 1997 Phospho-olivines as Positive-Electrode Materials for Rechargeable Lithium Batteries J. Electrochem. Soc. 144 1188–94

[8] Liu Y, Gorgutsa S, Santato C and Skorobogatiy M 2012 Flexible, Solid Electrolyte-Based Lithium Battery Composed of LiFePO4 Cathode and Li4Ti5O12 Anode for Applications in

Smart Textiles J. Electrochem. Soc. 159 A349
[9] Lee Y H, Kim J S, Noh J, Lee I, Kim H J, Choi S, Seo J, Jeon S, Kim T S, Lee J Y and Choi J W 2013 Wearable textile battery rechargeable by solar energy Nano Lett. 13 5753–61
[10] Dalton A B, Collins S, Razal J M and Howard V 2003 Super-tough carbon-nanotube fibres Nature 423 703
[11] Xu Z and Gao C 2011 Graphene chiral liquid crystals and macroscopic assembled fibres Nat. Commun. 2 571–9
[12] Vigolo B, Penicaud A, Coulon C, Sauder C, Pailler R, Journet C, Bernier P and Poulin P 2000 Macroscopic fibers and ribbons of oriented carbon nanotubes Science (80-.). 290 1331–4
[13] Zhao X, Zhenga B, Huanga T and Gao C 2010 Graphene-based single fiber supercapacitor with a coaxial structure Nanoscale 4 1166–9
[14] Chen Q, Meng Y, Hu C, Zhao Y, Shao H, Chen N and Qu L 2014 MnO2-modified hierarchical graphene fiber electrochemical supercapacitor J. Power Sources 247 32–9
[15] Aboutalebi S H, Jalili R, Esrafilzadeh D, Salari M, Gholamvand Z, Yamini S A, Konstantinov K, Shepherd R L, Chen J, Moulton S E, Innis P C, Minett A I, Razal J M and Wallace G G 2014 High-performance multifunctional graphene yarns: toward wearable all-carbon energy storage textiles ACS Nano 8 2456–66
[16] Zang X, Xu R, Zhang Y, Li X, Zhang L, Wei J, Wang K and Zhu H 2015 All carbon coaxial supercapacitors based on hollow carbon nanotube sleeve structure Nanotechnology 26 45401
[17] Gopalsamy K, Xu Z, Zheng B, Huang T, Kou L, Zhao X, Gao C and 5 2014 Bismuth Oxide Nanotubes/Graphene Fiber-Based Flexible Supercapacitors Nanoscale 6 8595–600
[18] Smithyman J and Liang R 2014 Flexible supercapacitor yarns with coaxial carbon nanotube network electrodes Mater. Sci. Eng. B Solid-State Mater. Adv. Technol. 184 34–43
[19] Sun W and Wang Y 2014 Graphene-based nanocomposite anodes for lithium-ion batteries. Nanoscale 6 11528–52
[20] Sun H, You X, Deng J, Chen X, Yang Z, Ren J and Peng H 2014 Novel graphene/carbon nanotube composite fibers for efficient wire-shaped miniature energy devices Adv. Mater. 26 2868–73
[21] Liu Y, Gorgutsa S, Santato C and Skorobogatiy M 2012 Flexible , Solid Electrolyte-Based Lithium Battery Composed of LiFePO 4 Cathode and Li 4 Ti 5 O 12 Anode for Applications in Smart Textiles J. Electrochem. Soc. 159 349–56
[22] Zhang Y, Zhao Y, Ren J, Weng W and Peng H 2016 Advances in Wearable Fiber-Shaped Lithium-Ion Batteries Adv. Mater. 28 4524–31
[23] Pampal E S, Stojanovska E, Simon B and Kilic A 2015 A review of nanofibrous structures in lithium ion batteries J. Power Sources 300 199–215
[24] Miao J, Miyauchi M, Simmons T J, Dordick J S and Linhardt R J 2010 Electrospinning of Nanomaterials and Applications in Electronic Components and Devices J. Nanosci. Nanotechnol. 10 5507–19
[25] Cavaliere S, Subianto S, Savych I, Jones D J and Rozière J 2011 Electrospinning: designed architectures for energy conversion and storage devices Energy Environ. Sci. 4 4761
[26] Gulzar U, Angelis F De, Zaccaria R P and Capiglia C 2016 Recent Advances in Textile Based Energy Storage Devices World J. Text. Eng. Technol. 2 6–15
[27] Miao J, Miyauchi M, Simmons T J, Dordick J S and Linhardt R J 2010 Electrospinning of nanomaterials and applications in electronic components and devices. J. Nanosci. Nanotechnol. 10 5507–19
[28] Leach M K, Feng Z-Q, Tuck S J and Corey J M 2011 Electrospinning Fundamentals: Optimizing Solution and Apparatus Parameters J. Vis. Exp. 2–6
[29] S R, K F, W T, T L and Z M 2005 An Introduction to Electrospinning and Nanofibers
[30] Toprakci O, Toprakci H A K, Ji L, Xu G, Lin Z and Zhang X 2012 Carbon nanotube-loaded electrospun LiFePO 4/carbon composite nanofibers as stable and binder-free cathodes for rechargeable lithium-ion batteries ACS Appl. Mater. Interfaces 4 1273–80
[31] Yarin A L, Koombhongse S and Reneker D H 2001 Taylor cone and jetting from liquid droplets in electrospinning of nanofibers J. Appl. Phys. 90 4836–46

[32] Khajavi R and Abbasipour M 2008 Controlling nanofiber morphology by the electrospinning process Electrospun Nanofibers vol 186pp 109–23
[33] Hohman M M, Shin M, Rutledge G and Brenner M P 2001 Electrospinning and electrically forced jets. II. Applications Phys. Fluids 13 2221–36
[34] Yuan X Y, Zhang Y Y, Dong C and Sheng J 2004 Morphology of ultrafine polysulfone fibers prepared by electrospinning Polym. Int. 53 1704–10
[35] Stanger J, Tucker N and Staiger M 1995 Electrospinning J. Electrostat. 75 218
[36] Frenot A and Chronakis I S 2003 Polymer nanofibers assembled by electrospinning Curr. Opin. Colloid Interface Sci. 8 64–75
[37] Homayoni H, Ravandi S A H and Valizadeh M 2009 Electrospinning of chitosan nanofibers: Processing optimization Carbohydr. Polym. 77 656–61
[38] Precursor P A N, By F and Technique E 2017 The Effect of Flow Rate , Concentration , and Voltage on Diameter of J. built Environ. Technol. Eng. 2 71–8
[39] Haider A, Haider S and Kang I K 2015 A comprehensive review summarizing the effect of electrospinning parameters and potential applications of nanofibers in biomedical and biotechnology Arab. J. Chem.

Ergonomic device to harvest Castilla-Blackberry

F Maradei[1], P Franco[1] and J M Castellanos[2]

[1]Universidad Industrial de Santander, Colombia
[2]Universidad Pontificia Bolivariana, Bucaramanga, Colombia

javier.castellanos@upb.edu.co

Abstract. Agriculture is one of the major economic activities in Colombia, counting for almost 47% of all country exportations. Considering Santander province, Castilla-Blackberry is one the most common agro-alimentary crops in the area (northeastern region of the country), but this particular cultivar presents thorns that rend harvesting difficult and increase the risk for skin or subcutaneous injuries. Other important issues of the harvesting processes are related to a rapid deterioration of the picking gloves, the homogeneous distribution of the fruit throughout the bush and the transportation of the harvested fruits during this activity (total load can be as high as 13 kg). In addition, the pinch gesture used to pick the fruit and the low back strain due to the heavy loads are responsible for hand pain and low back pain in this workers population. This work presents an experimental validation and a theoretical evaluation of a new ergonomic device to harvest Castilla-Blackberry, through a direct comparison with the traditional harvesting method. This new tool looks forward to reducing loads on the low back and to avoid cutaneous cuts. The experimental validation involved seven subjects in a real crop environment. Each participant, picked blackberry fruits in the morning hours, using both traditional elements and the new harvesting device. Meanwhile, posture data (articular angles variation) and hand cuts (gloves wear) were recorded. The new device supports up to 13 kg with a transfer system that reduces the intradiscal load in about 85.13%. Likewise, the pinch gesture was substituted with a supinate natural-hand closing gesture without detriment of the number of harvested fruits. Statistical analysis of trunk flexion does not evidence significant differences between both methods (p-value = 0.92), however, it was found a reduction in trunk flexion when the new device is used on sloping grounds. On the other hand, gloves wear analysis showed a decrease on the number of cuts and the washable materials used in the new device lead to a reduction of the annual use of new gloves in about 96.16%. In conclusion, the new device improves blackberry harvesting when biomechanical and physical risks are considered. Hence, it could be a viable alternative to be implemented in close collaboration with the concerned community, ensuring its appropriation of knowledge and promoting better labor conditions.

1. Introduction

Agriculture has been a worldwide engine that has allow overcome critical states of food shortage, poverty and unemployment, especially in developed countries. In Colombia, a positive expectation has been arisen because annual production of Castilla-Blackberry is about one hundred thousand tons per year [1]. Santander is a region located to the northeastern area of Colombia and it is the second larger blackberry producer of this country [2]. About 90% of the harvesting areas are mainly located in small towns like Piedecuesta, Floridablanca, Charta and Matanza [3]. On the other hand, there is evidence that agricultural activities can produce severe injuries due to postural behaviors, which are adopted by workers while performing harvesting activities. Most of the time, these behaviors cause musculoskeletal pain in back, shoulders, arms and hands; with a high prevalence [4–10]. This state of affairs suggests a worker population -normally smallholders- performing their tasks under painful conditions; unfortunately, 60% of this people are informal workers, which means that most of them do not have health care or security services. This situation exacerbates social problems and negatively compromises the sustainable development of Colombia [11]. Likewise, this country has troubling indicators of poverty for this population; 80% of them are smallholders with a very low annual income. Adding a long national social conflict to the overall picture, rural communities are vulnerable and exhibit inadequate standards of living [3].

The blackberry crop is manually harvested all around the Santander region because there are not suited tools to perform this activity. A preliminary work performed by one of the authors back in 2015, revealed that 72 workers in the Piedecuesta municipality exhibited musculoskeletal symptomatology and identified several contributing factors to this pathology. On one hand, features of this cultivar must be taken into consideration: irregular stems, presence of thorns and saw-toothed around the margins leafs rend difficult the harvesting task and increase the risk of skin or subcutaneous injuries. In addition, the pinch gesture to pick the fruit is performed uninterrupted [12]. Besides, this bush can grow as much as 3 meters in height and their fruits grow all over the plant. In consequence, the worker must adopt non neutral postures of the trunk and arms to reach the fruits at different zones (sometimes at the ground level or at the highest part of the bush) [12]. The last will explain the high prevalence reported in the preliminary work; 63.9% of participants alleged low back pain and 49.3% shoulders pain in the previous year.

Likewise, the review of the current technology showed that there are not manual tools for blackberry harvesting. The closest activity is harvesting of Brambleberry and Wild rose hip, where has been used a fruit picker that features a metallic or plastic comb [13]. This manual tool serves to pick and collect all fruits in a particular area at the same time. Afterwards, fruits are transferred to a larger container. Unfortunately, this device cannot be used with blackberries because the bush continuously grows fruits and the harvester must only pick the ripe ones and let the unripe in the plant [1].

On that basis, there is a clear need for technical solutions allowing to reduce pain perception and the possible occurrence of musculoskeletal disorders, while maintaining high levels of productivity. Blackberry harvesting is currently performed by rural communities on a weekly basis all year round, which means that activity demands are constant and repetitive during their entire working life. Consequently, this study looks forward to validating a new Castilla-Blackberry-harvesting system aiming to reduce not only loads on worker's back but also to avoid cutaneous cuts on his hands.

2. Materials and methods

To verify that the designed system allows reducing back loads and cutaneous cuts in workers, a single factor experiment (A), named "Harvesting system", was designed; this factor influences two variables: Back load (Y1) and cutaneous cuts (Y2).

2.1. Independent variable description

Factor A is composed by two treatments: "Traditional method" and "New harvesting system".

- Traditional method. Workers keep a plastic basket in front of their bodies, normally below the hips level, which is held using a riata or a rope on one shoulder and passing behind the back. This container stores the harvested fruit until is full, functioning as an intermediate stock system. The total weight of this set can be as high as 13 kg when the basket is full of fruits. Also, workers use black latex gloves in their hands to avoid being harm by the bush thorns (See Figure 1, left side).

- New harvesting system -SeCura-. The designed system is composed by an intermediate stock subsystem that distributes the load on both hips using two different deposits, hence dividing the total load on both legs. The system also features two gloves made of polymeric materials (See Figure 1, right side).

Figure 1. Traditional method to harvest Castilla-blackberry (left side) and SeCura system (right side)

2.2. Dependent variables description

The response to the load on the back (Y1) was studied using a biomechanical model of the back for every treatment (Traditional method and SeCura). On first place, the average back flexion was computed for every participant as a function of every treatment. Afterwards, the general average flexion was calculated for every treatment.

Likewise, response Y2 -i.e. cutaneous cuts- was studied by means of the gloves-material wear. Every participant used new gloves at the beginning of the experiment in both treatments.

2.3. Experiment description

Table 1 exhibits the demographic description of 72 Castilla-blackberry harvesters, working in the rural area of Piedecuesta Municipality. The procedure used to determine these anthropometrical data is described in another study of the northeast-region population of Santander [14]. Hence, this work involved participants with the same somatometric features as it is shown in Table 2.

Table 1. Demographic description of Castilla-Blackberry harvesters in Piedecuesta, Santander (2015)

Workers Population	Gender	Height (m)	Weight (kg)	BMI	Dominant hand
N=72	55.6% male 55.4% female	1.6	66.3	24.31	97% right 3% left

Table 2. Demographic description of the participants on this study (April, 2017)

Study participants	Gender	Height (m)	Weight (kg)	BMI	Dominant hand
	M	1.7	75	25.95	left
	M	1.7	95	32.87	Left
	F	1.67	55	19.72	right
N=7	F	1.68	70	24.80	right
	F	1.64	60	22.31	right
	F	1.58	52	20.83	right
	M	1.7	62	21.45	right
Total	40% male 60% female	1.667	67	24	72% right 28.5% left

As a first step, each participant was informed about the details of the experiment, declaring as well that enrolment in the study was voluntary. Then, participants gave their authorization to make the experiment and to collect data. Immediately before the harvesting activity begins, optical markers were installed to indicate the following points in the sagittal view: Greater trochanter, iliac, acromial, radial and styloid ulnar (See figure 2). In the posterior view, optical markers were installed on the L5 and T12 vertebrae.

The experimentation was performed on a real Castilla-blackberry plantation under real conditions; a video camera was used for recording the harvesting activity in both treatments, while time employed in every test was fixed on three hours.

Flexion and extension of the back were determined using the articular angles indicated by the optical markers by means of an image processing procedure in the Kinovea® Software. Three hours of video per participant were analyzed with a frequency of three frames per second. The load on the back -i.e. on L5-S1 vertebrae- was calculated using the Biomechanical model proposed by Irving Herman [15], which considers that the force exerted by the erector muscle forms a 12 degrees angle with the longitudinal axe of the column (See figure 3). To compute the intradiscal force at different positions during the harvesting process, it was considered that the maximal load of the basket- serving as intermediate stock in the traditional method- was 13 kg. Besides, this weight is supported by the shoulder. On the contrary, there is no load transferred to the back when using SeCura system.

3. Results

3.1 Flexion and extension of the back

Descriptive analyses of data show that flexion and extension angles of the back are lower on SeCura, however, there are not significant differences in the flexion angle (p-value = 0.92) or in the extension angle (p-value=0.07). It is important to outline that cero degrees extension and flexion angles were computed with SeCura System, which yields higher standard deviations than the traditional method (See Table 3).

IOP Conf. Series: Materials Science and Engineering **437** (2018) 012007 doi:10.1088/1757-899X/437/1/012007

Figure 2. Visual markers location and back flexion angle on one video frame

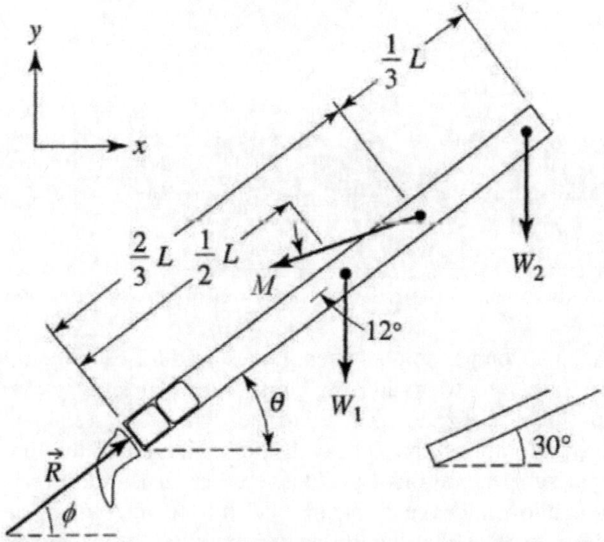

Figure 3. Biomechanical model used to compute intradiscal loads

Table 3. Comparative results for back flexion and extension angles in both treatments (N=7)

	Traditional Method		SeCura	
	Flexion (°)	Extension (°)	Flexion (°)	Extension (°)
Mean Values	14.99	6.15	14.23	2.62
Maximal value	29.12	9.14	48.6	9.24
Minimal value	4.83	2	0	0
Standard deviation (SD)	8.3	3.1	18.0	3.6

IOP Conf. Series: Materials Science and Engineering **437** (2018) 012007 doi:10.1088/1757-899X/437/1/012007

3.2. Intradiscal load on L5-S1 vertebrae

Table 4 shows that intradiscal load on L5-S1vertebrae is reduced in about 85.13% when harvesters use SeCura system. Likewise, the study reveals significant differences (p-value=0.002) between traditional method and SeCura.

Table 4. Comparative results for intradiscal loads in both treatments (N=7)

	L5-S1 Intradiscal load (N)	
	Traditional Method (13 kg on the back)	SeCura (No load on the back)
Mean values	2818.69	419.06
Maximal value	5353.73	1357.7
Minimal value	926.31	0.0
Standard deviation (SD)	1508.81	509.93

3.3. Wear of gloves material

Cutaneous cuts were evaluated using gloves wear after three hours of harvesting with both methods.

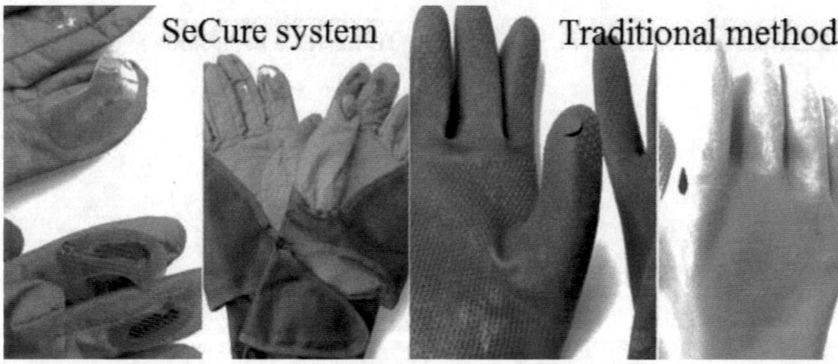

Figure 4. Comparison between glove wear for both treatments (SeCure and Traditional method)

Figure 4 shows a graphical comparison between gloves used for each treatment (Traditional method and SeCure). It can be observed wear and big perforations in latex gloves (Traditional method), especially on fingers-tips and dorsal zones. Castilla-blackberry harvesters usually change these gloves every 15 days; however, perforations can appear at the beginning of this interval. On the contrary, SeCure gloves did not undergo perforations and cuts were minimal. Additionally, they have been made of washable materials that allow their reuse for long periods of time. Considering that SeCure gloves are changed once a year, the estimated reduction on current gloves replacement rate is 96.16%

4. Discussion and conclusions

This work looked forward to evidence whether SeCura system -designed as a tool to harvest Castilla-Blackberry- was able to reduce back load as well as cutaneous cuts in harvesters. According to the ergonomic analysis of the blackberry harvesting activity, there is evidence of musculoskeletal loads in this particular population. Their causes are related to the heavy weight of the basket used as intermediate stock in the traditional method, as well as the presence of thorns all over the blackberry cultivar.

Several studies show that there is strong evidence that lifting of heavy loads and performing movements with loads are associated with an increased risk of low back pain [16]; this pathology occurs due to the loads on the musculoskeletal system and the adopted postures[17]. In consequence, when the intermediate stock container is no longer supported by the back, loads on the trunk-musculoskeletal system are -as it is shown in this study- immediately reduced.

The designed system avoids additional loads to the back because the harvested fruit and the container weights are directly transferred to the worker´s heaps; this allows to reduce the intradiscal load in about 85.13% when it is compared to the traditional system.

In relation with adopted postures -i.e. flexion and extension of the back- influencing the presence of low back pain[16], this study did not found differences between both treatments (p-value=9.92). Therefore, the weight of the intermediate stock system in the traditional harvesting method and the necessary muscles coactivity to keep equilibrium and support the load [18] are the factors that directly influence the intradiscal load.

Likewise, literature indicates that the maximal load supported by the L5-S1 articulation is 3400N [19]; this study found heavy loads –near to this maximal value- when the traditional method is employed (2819.9N on average). The results obtained agree well with literature, which indicates that a 10 kg load produces internal loads between 2820N and 7260N[20]. Nonetheless, loads computed with SeCura system were 419.06N on average.

Finally, the designed system offers a long-duration reusable glove; when compared with latex gloves use, the annual rotation of the former is 96.16% lower that the latter. This contributes to reduce the environmental impact of wasted latex gloves. Besides, substitution of latex gloves can contribute to reduce other skin pathologies. According to literature, latex contact produces dermatitis and the allergic prevalence to this material is 1% [21]. Many harvesters acknowledge that they do not use this kind of gloves due to this allergic pathology, situation that increases the risk of cutaneous cuts and biological contamination of the harvested fruits.

References

[1] Rativa C M G, Medellín L A C and Trujillo M M P 2016 Rendimiento y calidad de la fruta en mora de Castilla (Rubus glaucus Benth), con y sin espinas, cultivada en campo abierto en Cajicá (Cundinamarca, Colombia) *Revista Facultad de Ciencias Básicas* **6** 24–41

[2] Betancurt E, García E, Giraldo M, Quejada O, Rodríguez H and Arroyave I 2014 Manual técnico del cultivo de maracuyá bajo buenas prácticas agrícolas *Servicio Nacional de Aprendizaje (SENA). Medellin-Colombia*

[3] Ministerio de Agricultura y Desarrollo Rural 2006 *Desarrollo de la fruticultura en Santander*

[4] Adams M A, McMillan D W, Green T P and Dolan P 1996 Sustained Loading Generates Stress Concentrations in Lumbar Intervertebral Discs *Spine* **21** 434

[5] Rejoice S, Dina. A and Arhin A 2013 Occupational health hazards and safety of the informal sector in the metropolitan area of ghana *Research on humanities and social science* **3** 87–99

[6] Grzywacz J G, Arcury T A, Mora D, Anderson A M, Chen H, Rosenbaum D A, Schulz M R and Quandt S A 2012 Work organization and musculoskeletal health: clinical findings from immigrant Latino poultry processing and other manual workers *Journal of occupational and environmental medicine* **54** 995–1001

[7] Hagberg M and Wegman D 1987 Prevalence rates and odds ratios of shoulder-neck diseases in different occupational groups. *Occupational and Environmental Medicine* **44** 602–610

[8] Norman R, Wells R, Neumann P, Frank J, Shannon H, Kerr M and others 1998 A comparison of peak vs cumulative physical work exposure risk factors for the reporting of low back pain in the automotive industry *Clinical biomechanics* **13** 561–573

[9] Pal P, Milosavljevic S, Gregory D E, Carman A B and Callaghan J P 2010 The influence of skill and low back pain on trunk postures and low back loads of shearers *Ergonomics* **53** 65–73

[10] Silverstein B A, Bao S S, Fan Z J, Howard N, Smith C, Spielholz P, Bonauto D and Viikari-Juntura E 2008 Rotator cuff syndrome: personal, work-related psychosocial and physical load factors *Journal of occupational and environmental medicine* **50** 1062–1076

[11] Semana 2012 Los pilares olvidados de la tierra *Revista Semana*

[12] Fathallah F A 2010 Musculoskeletal disorders in labor-intensive agriculture *Applied ergonomics* **41** 738–743

[13] Muñoz O and Cattan M 2007 Equipamiento para la recolección de rosa mosqueta silvestre

[14] Maradei F, Correal F E and Peña A A 2008 Estudio de valores antropométricos para la región nororiental colombiana. *Revista UIS Ingenierías* **7** 153–167

[15] Irving P H 2007 *Physics of the human body* (springer)

[16] Marras W S, Lavender S A, Leurgans S E, Rajulu S L, Allread S W G, Fathallah F A and Ferguson S A 1993 The Role of Dynamic Three-Dimensional Trunk Motion in Occupationally-Related *Spine* **18** 617–628

[17] Bernard B P, Health N I for O S and and Prevention C for D C and 1997 Musculoskeletal disorders and workplace factors: a critical review of epidemiologic evidence for work-related musculoskeletal disorders of the neck, upper extremity, and low back *Musculoskeletal disorders and workplace factors: a critical review of epidemiologic evidence for work-related musculoskeletal disorders of the neck, upper extremity, and low back*

[18] van Dieën J H, Dekkers J J, Groen V, Toussaint H M and Meijer O G 2001 Within-subject variability in low back load in a repetitively performed, mildly constrained lifting task *Spine* **26** 1799–1804

[19] Arjmand N, Amini M, Shirazi-Adl A, Plamondon A and Parnianpour M 2015 Revised NIOSH Lifting Equation May generate spine loads exceeding recommended limits *International Journal of Industrial Ergonomics* **47** 1–8

[20] González Jemio F, Mustafá Milán O and Antezana Arzabe A 2011 Alteraciones biomecánicas articulares en la obesidad *Gaceta Médica Boliviana* **34** 52–56

[21] Gaspar Carreño M, Arias Pou P, Rodríguez Berges O, Gamundi M and Carbonell Tatay F 2011 Revisión sobre el uso de guantes en los hospitales. *El Farmacéutico Hospitales*

IOP Conf. Series: Materials Science and Engineering **437** (2018) 012008 doi:10.1088/1757-899X/437/1/012008

Estimation of energy efficiency in solar photovoltaic panels considering environmental variables

R Martinez[1] **and E Forero**[1]

[1]Faculty of Electronics Engineering, Universidad Santo Tomás, Kr 9 No. 51 – 11, Bogotá, Colombia

edwinforero@usantotomas.edu.co

Abstract. The present work has been made to find out the relationship between the efficiency of two types of photovoltaics panels, monocrystalline and polycrystalline, and the weather variation in the city of Bogota-Colombia. The power generated of two photovoltaic panels, one monocrystalline and one polycrystalline, were collected for one month, 24 hours in condition of variant weather. At the same time, it was implemented a weather station that registered the variations of climatic variables like wind speed, temperature, humidity, irradiance and rainfall, to relate them with the generated power of each photovoltaic panel. Once with this information collected it a study was made with all the correlated variables through of the statistic tool PCA (Principals Components Analysis) where it was seen the impact of each climatic variable over the generated power in both panels and therefore its impact over the efficiency.

1. Introduction

Due to the great growth that has had the use of solar energy in the world, the industry has evolved in materials and manufacturing technics of solar panels to lower production costs. However, the efficiency of the photovoltaic panels is still low, up to 17% in laboratory tests, being also affected by climatic factors of the region where the panels are installed [1].

For this reason, it is important to stablish the efficiency of the photovoltaic panels in real conditions of installation, since efficiency giving by manufacturers is established in laboratory conditions what can vary significantly in field by climatic factors such as temperature, wind, rainfall, humidity and solar radiation [2].

This document presents the implementation of a weather station that registered the climatic variables that affect the generation of solar energy in the city of Bogota and the generated power by two photovoltaic solar panels, one of monocrystalline type and other of polycrystalline type, to do an analysis through a statistical tool called PCA (principal component analysis) on how the climatic variables are correlated with the power obtained as well as the variation in the efficiency of solar panels with the climatic conditions registered.

2. Efficiency in photovoltaic solar panels

IOP Conf. Series: Materials Science and Engineering **437** (2018) 012008 doi:10.1088/1757-899X/437/1/012008

Solar energy can be transformed into electricity using the photovoltaic effect, which increase the electrical conductivity of a material under the action of light, due to the generation of charge carriers and holes [3].

2.1. Photovoltaic cell

A photovoltaic cell is a device that directly converts light into electric energy based on the photovoltaic effect [4]. Most photovoltaic cells are made of silicon. A photovoltaic cell consists of a PN junction, two electrodes and a conductive network anti reflection. A solar cell is basically made of two semiconductors, the first N-type and the second P-type. The N-type semiconductor is the component of the cell that will receive the light of the Sun, while P-type semiconductor will be on the back of the solar cell. The union of both kinds of semiconductors made up the junction PN [5].

2.2. Photovoltaic panels and Irradiance

The types of solar cells made from silicon are monocrystalline, polycrystalline and amorphous. Monocrystalline silicon cells are the most efficient, up to 17%, although the manufacturing process is expensive and has a deficient performance in conditions of poor lighting. The cells are cut from a silicon crystal, are rigid and must be mounted in a rigid framework to protect them. Polycrystalline silicon cells have an efficiency up to 13% and a low cost of production. These cells are cut from a silicon block consisting of many crystals.

Irradiance is the unit used to describe the incident solar radiation per unit area and is measured in W/m^2. In a photovoltaic module, the incidence of solar radiation causes the appearance of generated current that is directly proportional to the irradiance. In short circuit operation, the resulting current varies in proportion to irradiance [6].

2.3. Efficiency panels solar

The efficiency of a solar cell is given by the ratio between the power output of the cell Pout and the incident power Pin, where Pin is the product of irradiance and the area of the cell. In (1) the efficiency of the cell parameters given by the manufacturer is related, as maximum power (Pmax), maximum voltage power (Vmp), maximum current power (Imp), open circuit voltage (Voc), and short circuit current (Isc).

$$\eta = \frac{\mathrm{Pout}}{\mathrm{Pin}} = \frac{\mathrm{Pmax}}{\mathrm{Pin}} = \frac{\mathrm{ImpVmp}}{\mathrm{Pin}} = \frac{\mathrm{IscVocFF}}{\mathrm{Pin}} \quad (1)$$

$$\mathrm{FF} = \frac{\mathrm{Pmax}}{\mathrm{IscVoc}} \quad (2)$$

In (2) is related the fill factor (FF), which is the ratio of the maximum power and the product between short circuit current (Isc) and open circuit voltage (Voc) [6].

2.4. PCA principal components analysis

To analyze the impact that have the climatic variables have on the power of photovoltaic panels, it was used a multivariate statistical technique called Principal Components Analysis PCA. The PCA is a statistical technique that takes data and transforms it in such a way that new data have statistical properties. The statistical properties are chosen so that the transformation highlights the importance of the data elements. The data can also be reduced or compressed by removing or filtering elements less important [7]. The PCA is based on the search of linear combinations of elements of a vector X to find a combination with the largest variance. The principal components are derived from eigenvalues and eigenvectors of the matrix of covariance VarX [8].

IOP Conf. Series: Materials Science and Engineering **437** (2018) 012008 doi:10.1088/1757-899X/437/1/012008

The PCA is also known as a technique for exploratory data analysis that basically projected data in a reduced hyperspace defined by orthogonal principal components. These components are linear combinations of original variables, where the first principal component has the largest variance, the second component has the second variance and thus the other components. Therefore, it is possible to select a considerable number of components in such a way that the dimension of the data is reduced while preserving the systematic variation in the data retained on the first principal components, while the noise is represented in the latest components [9].

3. Station weather and power measurement

To carry out the research on the correlation of climatic factors with the efficiency of solar photovoltaic panels, it was designed a weather station that allowed capture and record both climatic variables and power generated by the solar panels. This station is made up of a solar radiation sensor, a temperature sensor, an anemometer, a rain collector, a humidity sensor and a power interface (Figure 1).

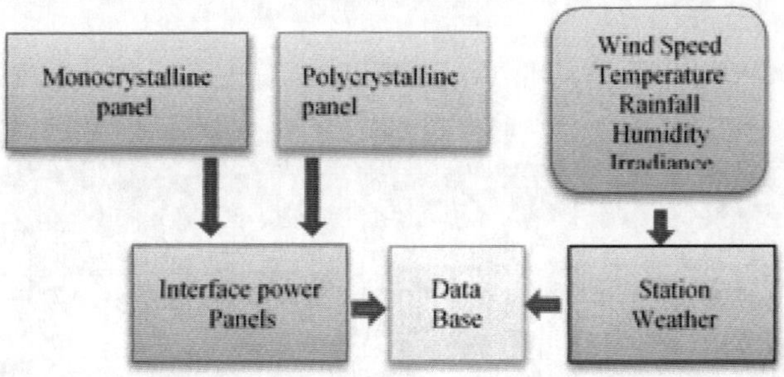

Figure 1. Design of weather station

3.1. Solar radiation Sensor

A Pyranometer that measures both the direct and diffuse components of solar irradiance and converts the incident radiation into proportional voltage was used for the measurement of the irradiance. The transducer is a photodiode hermetically sealed that includes an amplifier that converts the transducer current at 0 to 2.5 volts. The analog output is of 1.67 mv per W/m², and its conversion is made at the station through an ADC channel (Figure2).

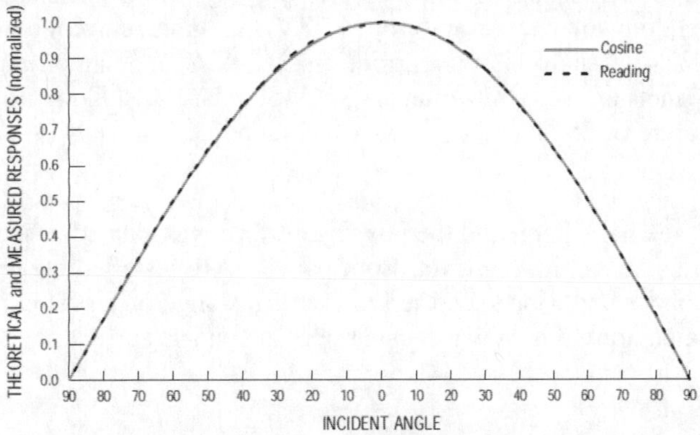

Figure 2. Response solar radiation sensor

3.2. Power measurement interface

To measure the power generated it was designed an interface connected to two solar panels with similar characteristics (50W@12V) but different technology (monocrystalline and polycrystalline), where it was measured as the voltage as the current in the output of each panel (Figure 3).

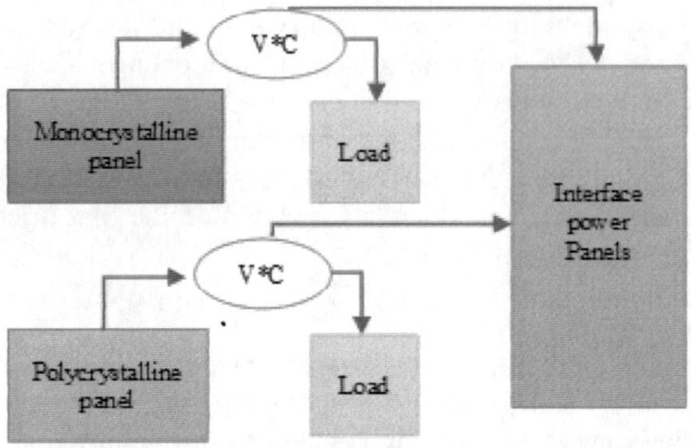

Figure 3. Interface of Power measurement

3.3. Solar panels

Figure 4. Monocrystalline (left) and polycrystalline (right) panels

The monocrystalline panel has a maximum power (Pmax) of 50 watts, maximum voltage power (Vmp) 18.5V, maximum current power (Imp) 2.70A, open circuit voltage (Voc) 22.7 V, short circuit current (Isc) 2.84A and an efficiency of 14.67%. The polycrystalline panel has a maximum power (Pmax) of 50 watts, maximum voltage power (Vmp) 17.8V, maximum current power (Imp) 2.80A, open circuit voltage (Voc) 22.4V, short circuit current (Isc) 2.95A and an efficiency of 13.37%. All specifications in these panels are determined under STC (Standard Test Conditions) with an irradiance of 1000W/m², a temperature of 25 °C and air mass 1.5 (Figure 4).

3.4. Implementation

The implementation of weather station and the power interface was carried out in the city of Bogotá at an altitude of 2592 meters above the sea level. Both panels were installed under the same conditions. Rainfall, humidity, irradiance and wind speed sensors were installed on a post parallel to the location of the panels, while the temperature sensor was located near the panels surface.

4. Testing and results

Once the weather station and the power interface were implemented and validated, all the variables were registered every two minutes since October 20 to November 20, 2016. The records were stored in a log file that included the timestamp, wind speed(m/s), rainfall (mm), irradiance (W/m²), relative humidity (%), temperature (°C), voltage (V) and current (mA) of each panel. The first analysis began with some graphics showing the relationship between climatic variables and the irradiance to observe the impact that may have on the power generated by the panels. For this analysis were selected different days where some climatic conditions were more relevant than others.

4.1. Irradiance and temperature

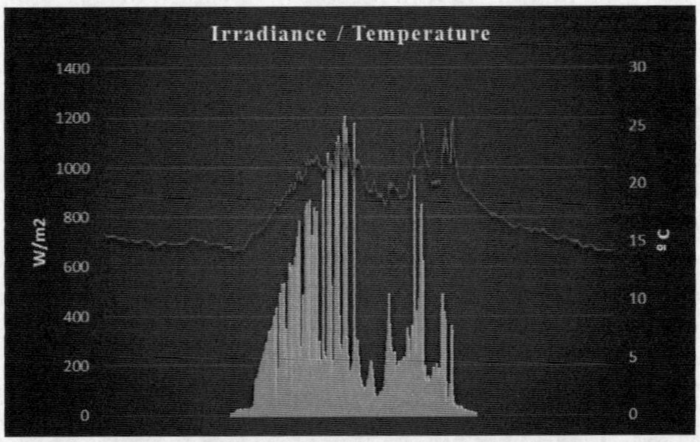

Figure 5. Irradiance and temperature

The figure 5 corresponds to records of October 20, where there were conditions of dry weather for good levels of irradiance and a warm temperature. The graphic shows that irradiance levels behave according to the temperature levels. The left side of the graphic shows the irradiance scale in W/m², where is seen that this variable reaches the 1200W/m². The the right side shows the temperature scale in °C, where is seen that this variable reaches a peak of 25 °C when the irradiance levels are the highest. The highest levels of irradiance occurred around 11:00 am.

4.2. Irradiance and rainfall

The figure 6 corresponds to records on October 26, where there was intense rainfall producing low levels of irradiance.

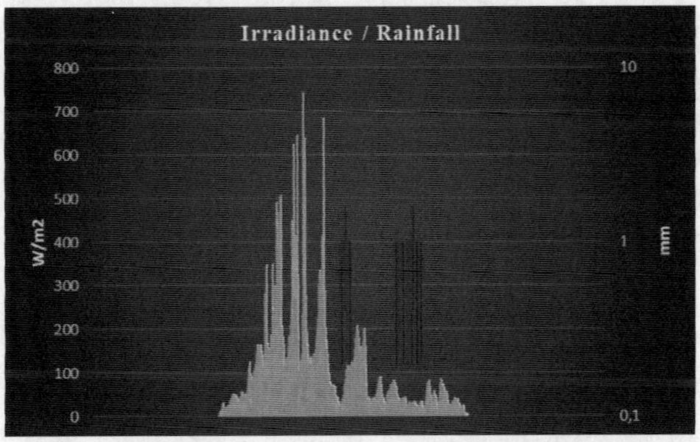

Figure 6. Irradiance and rainfall

The graphic shows that irradiance levels fall dramatically with the rainfall. The left side of the graphic shows the irradiance scale in W/m² which reaches 750 W/m². On the right the rainfall is showed achieving levels of 1.6 mm in just 2 minutes. On this day, the weather station records a total of 13.8 mm of rainfall.

4.3. Irradiance and relative humidity

The figure 7 corresponds to records on October 27 where the weather conditions were cloudy producing low levels of irradiance and elevated levels of humidity. The graphic shows that irradiance levels rise when the humidity levels fall. The left side of the graphic is the irradiance scale in W/m², which reaches 1000 W/m². The right side is the humidity, which achieves maximum levels of 90% and minimum of 41.2%.

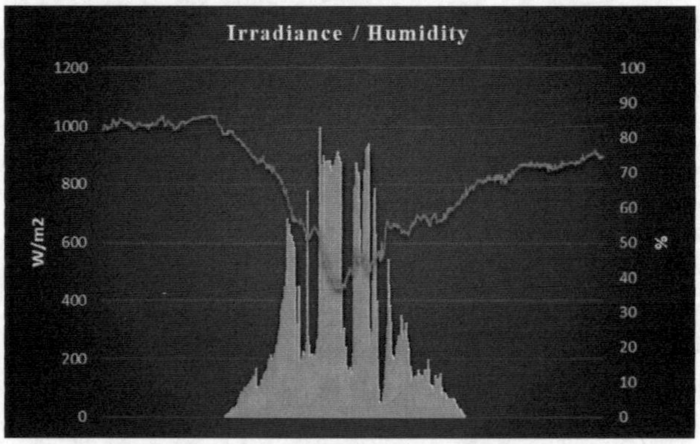

Figure 7. Irradiance and humidity

4.4. Irradiance and wind speed

The figure 8 corresponds to records on October 22, where there were conditions of cloudy weather with intense winds.

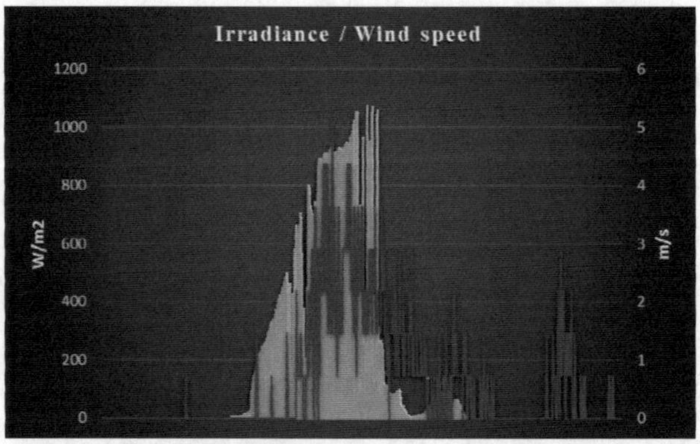

Figure 8. Irradiance and wind speed

The graphic shows that the irradiance levels rise when the wind speed increases, however is not clear a direct correlation between these variables. The left side of the graphic shows the scale of irradiance in W/m², which achieves 1075 W/m². The right is the wind speed scale which achieves maximum levels of 5.1 m/s.

4.5. Power generated by the panels according to the irradiance

The figure 9 corresponds to records on October 29, where there was cloudy weather with persistent rainfall.

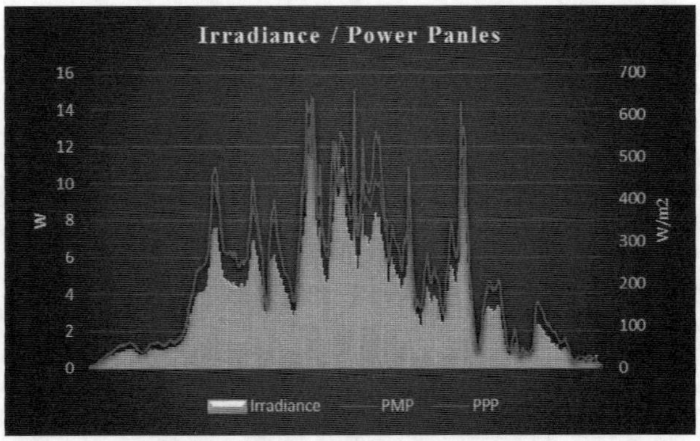

Figure 9. Power panels and irradiance

The graphic compares the power generated by the photovoltaic panels (Power of Polycrystalline Panel PPP and Power of Monocrystalline Panel PMP) with incident irradiance. The graphic shows that the levels of power of the two panels follow the same pattern of irradiance. It is seen that monocrystalline panel delivers more power at the maximum values of irradiance than the polycrystalline panel. On the right, the graphic shows the irradiance scale, which reaches maximum 600 W/m². On the left is the power panels scale, where it is seen that the panels achieve a 15 watts of output power.

4.6. Monocrystalline Panel Power and temperature

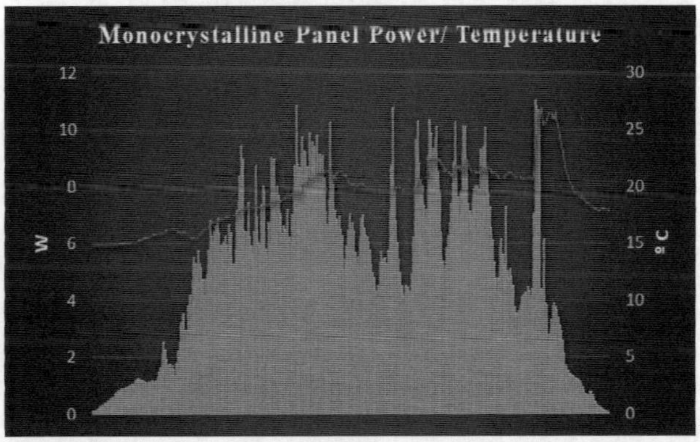

Figure 10. Monocrystalline Panel Power and temperature

The figure 10 corresponds to records on October 30, where there was dry weather with a maximum temperature of 25° C. In the graphic is seen the relationship between the temperature and the generated power of the monocrystalline panel, where the maximum temperature levels correspond with the panel power. The left side of the graphic shows the power scale, which achieves the 11W. On the right, there is the temperature scale which achieves a maximum level of 25 °C.

4.7. Monocrystalline Panel Power and wind speed

The figure 11 corresponds to records on October 30, where there were weather conditions with recurrent winds. In the graphic a direct relationship between the wind speed and the panel power is not seen. The left side of the graphic shows the panel power scale, which reaches until 11W. On the right, is seen that the wind speed scale gets a speed of 3.64 m/s.

IOP Conf. Series: Materials Science and Engineering **437** (2018) 012008 doi:10.1088/1757-899X/437/1/012008

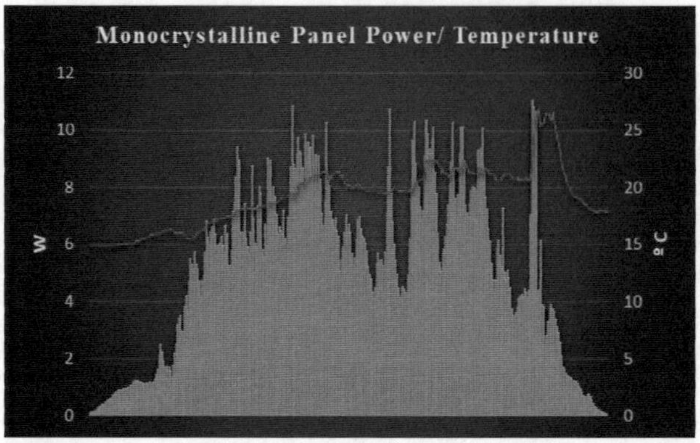

Figure 11. Monocrystalline Panel Power and Wind Speed

4.8. Monocrystalline Panel Power and rainfall

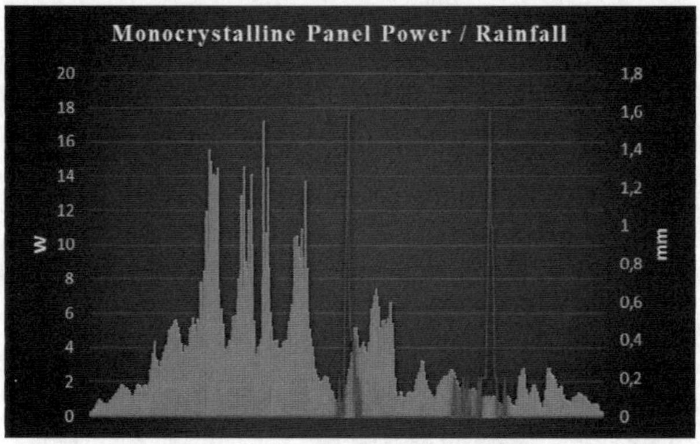

Figure 12. Monocrystalline Panel Power and rainfall

The figure 12 corresponds to records on October 26 where there was cloudy weather with intense rainfall. In the figure 12 is seen the relationship between rainfall and the generated power of the monocrystalline panel, where maximum levels of rainfall correspond to low levels of the panel power. The left side of the graphic shows th panel power scale, which achieves the 17 W. On the right, it is showed the rainfall scale that achieves maximum levels of 1.6 mm.

4.9. Monocrystalline Panel Power and humidity

The figure 13 corresponds to records on November 2, where there was a cloudy weather. In the graphic is seen the relationship between the humidity and the power generated for the monocrystalline panel, where the maximum humidity levels coincide with the lowest levels of the power panel. The left side of the graphic shows the panel power scale, which achieves the 19W. On the right, the humidity scale is showed achieving the minimum level of 58%.

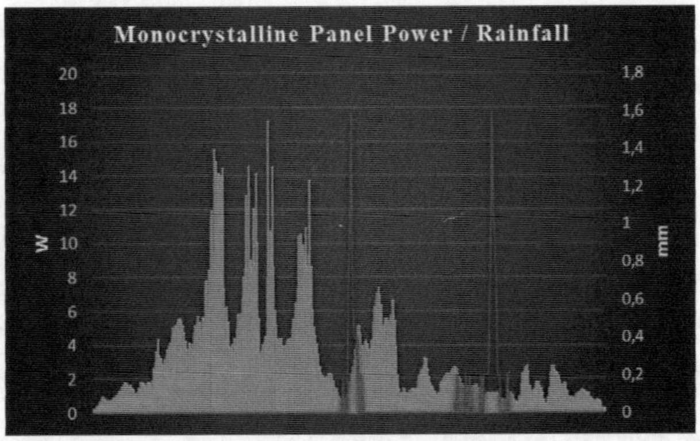

Figure 13. Monocrystalline Panel Power and relative humidity

5. Analysis of results

Once seen the relationship among the different climatic variables (temperature, humidity, rainfall and wind speed) and the irradiance through the graphics and after a preliminary analysis of the impact that have these variables have on the panels power, it was made a deeper analysis based on the information collected by the weather station. From the information obtained of the irradiance and the power output of the solar panels, a polynomial regression determines a function that relates both variables. In (3) is shown the function obtained from the polynomial regression according to the irradiance and the power panel.

$$y = -10^{-5}x^2 + 0{,}0339x - 0{,}2681 \quad (3)$$

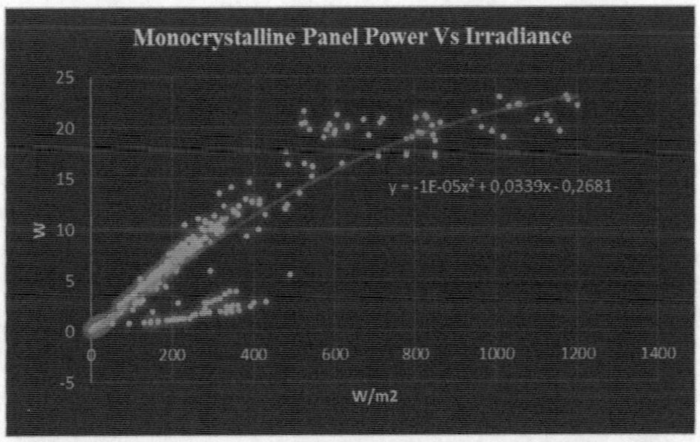

Figure 14. Scatter plot Panel Power Vs irradiance

In the coefficient matrix shown in the figure 15, the variables that have a positive correlation are presented in blue color, while those that have negative correlation are presented in red colour. More tenuous colours indicate that the variables are not correlated. For example, the irradiance has a negative correlation with the humidity, while it has a positive correlation with the current.

Additionally, it was obtained an array of scatter plots (figure 16) between every variable to appreciate how they relate each other, however as the correlation coefficients matrix as the scatter plot matrix only allows to see the relationship between pairs of variables. For this reason, it was used a multivariate statistical tool called PCA (principal component analysis) to be able to determine how every variable is correlated with the power obtained from the photovoltaics panels.

Pearson's r	Wind	Rain	Irrad	humid	Temp	VP	CP
Wind	-	0,450	0,593	-0,155	0,589	0,780	0,644
Rain	0,450	-	-0,242	0,333	-0,048	-0,102	-0,197
Irrad	0,593	-0,242	-	-0,558	0,613	0,867	0,948
Humid	-0,155	0,333	-0,558	-	-0,805	-0,535	-0,522
Temp	0,589	-0,048	0,613	-0,805	-	0,817	0,639
VP	0,780	-0,102	0,867	-0,535	0,817	-	0,910
CP	0,644	-0,197	0,948	-0,522	0,639	0,910	-

Figure 15. Correlation coefficients Matrix

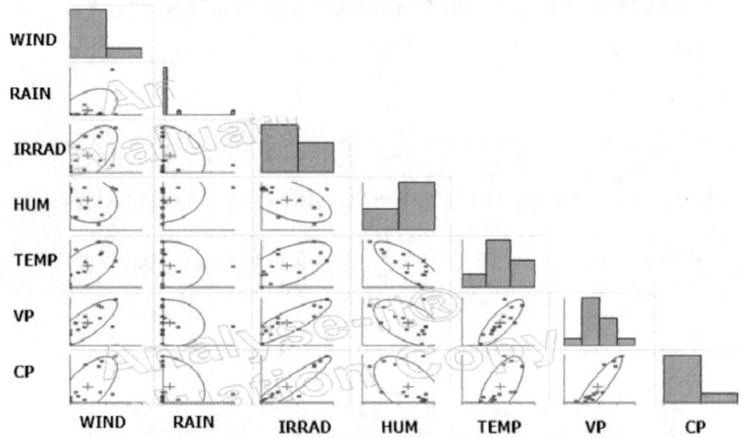

Figure 16. Scatter Plot Matrix

To make a PCA analysis, it is necessary to select the observers, in this case the records of time where the variables were correlated. Since the magnitudes of the variables are very different, the information contained was normalized in the matrix by dividing each column by its standard deviation, in this way the information contained in the matrix is made dimensionless and the significant differences between the variances of the variables decrease.

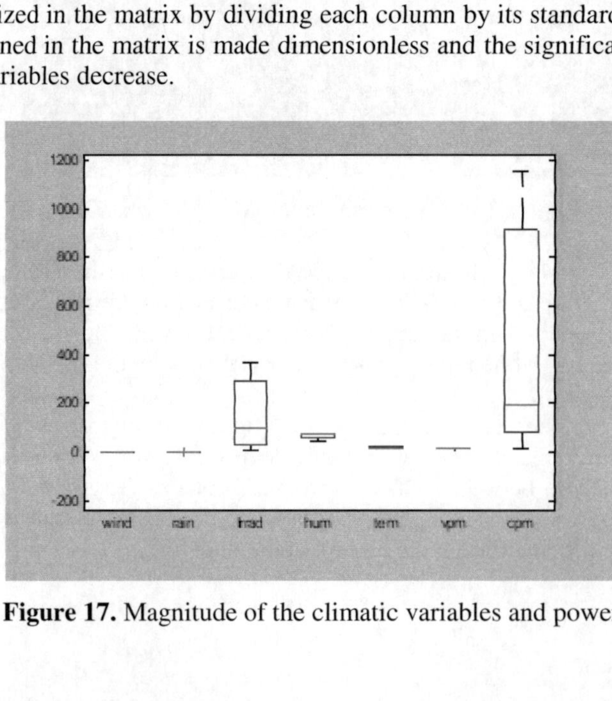

Figure 17. Magnitude of the climatic variables and power

5.1. PCA Monocrystalline Panel Power and irradiance

Table 1. PCA - Irradiance coefficients

	PC1	PC2	PC3	PC4	PC5	PC6	PC7
WIND.	0,35	-0,41	0,793	0,223	-0,140	0,045	0
RAIN.	0	0	0	0	0	0	1
IRRAD.	0,38	0,566	-0,048	0,717	0,095	-0,08	0
HUME.	-0,41	0,404	0,372	0,003	0,101	0,717	0
TEMP.	0,429	-0,28	-0,447	0,084	-0,262	0,677	0
VOL.P.	0,464	0,038	0,052	-0,366	0,795	0,109	0
COR.P	0,398	0,508	0,159	-0,542	-0,508	-0,06	0

The first principal component analysis is done in dry weather with elevated levels of irradiance. In this first analysis PCA were taken as the environmental variables (wind speed, rainfall, irradiance, humidity and temperature) and the power panels variables (voltage panel VP and current Panel CP). To determine which are the most principal components a vector called latent is found, which indicates the amount of accumulated variance provided for each principal component.

Table 2. Variance components

	PC1	PC2	PC3	PC4	PC5	PC6	PC7
Latent	0.74	0.910	0.983	0.994	0.998	1.000	1.000

According to latent vector, it is seen that the components that are most important are the first two, which reached 91% of the variance. In the table of coefficients is seen that PC1 has the highest positive correlation with the variables of power panel (voltage and current), temperature and irradiance, while it has negative correlation with humidity. PC2 shows correlation positive between the variables of irradiance and current while negative correlation with the wind speed. Since there was not rainfall on this day its coefficient is zero.

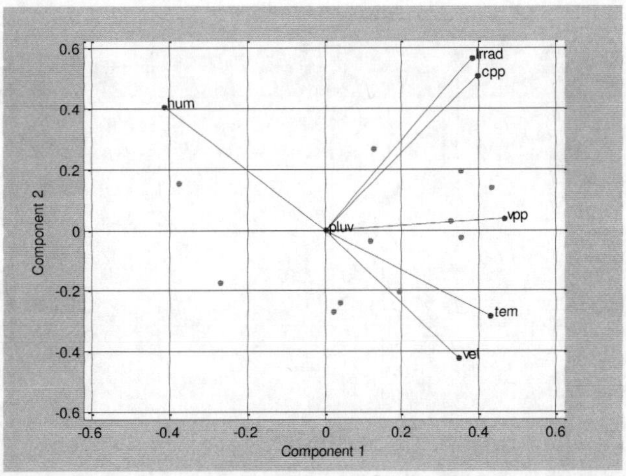

Figure 18. Monocrystalline Panel main components - irradiance

Since the two first principal components have the greater part of the variance, all variables are plotted according these components. In Figure 18 the correlation between all the variables with the power output of the panel is clearly seen, the vectors of variables that are found with a small angle with respect to irradiance have a positive correlation, while those that have approached 90 degrees angle have a negative

IOP Conf. Series: Materials Science and Engineering **437** (2018) 012008 doi:10.1088/1757-899X/437/1/012008

correlation, as it is the case of humidity also shows that the rainfall has not a vector since that was not present in the samples.

5.2. PCA Monocrystalline Panel and rainfall

The second principal component analysis is done with persistent rainfall and cloudy weather conditions. In this analysis PCA were taken as the environmental variables (wind speed, rainfall, irradiance, humidity and temperature) as the power panels variables (voltage panel VP and current Panel CP).

Table 3. PCA - Rainfall coefficients

	PC1	PC2	PC3	PC4	PC5	PC6	PC7
WIND.	0,1962	0,6185	-0,328	0,2791	-0,533	0,32	0,04
RAIN	-0,086	0,6148	0,7559	-0,003	0,2074	-0,0	0,00
IRRAD.	0,4582	-0,165	0,2688	-0,247	-0,334	-0,1	0,70
HUME.	-0,455	-0,166	0,1346	-0,355	-0,152	0,75	0,14
TEMP.	0,3973	0,2932	-0,342	-0,543	0,5456	0,21	0,04
VOL.P.	0,4134	-0,267	0,1784	0,5949	0,351	0,49	0,06
COR.P	0,4557	-0,162	0,2854	-0,293	-0,341	0,11	-0,6

According to the latent vector, it is seen that the most important components are the first two, which reached 87% of the variance.

Table 4. Variance components

	PC1	PC2	PC3	PC4	PC5	PC6	PC7
Latent	0.602	0.877	0.936	0.979	0.994	0.999	1.000

In the table of coefficients is seen that PC1 has the highest positive correlation with the variables of power panel (voltage and current) and irradiance, while it has negative correlation with humidity. PC2 shows a positive correlation between the variables of wind speed and rainfall since it was a day where there was an intense rain.

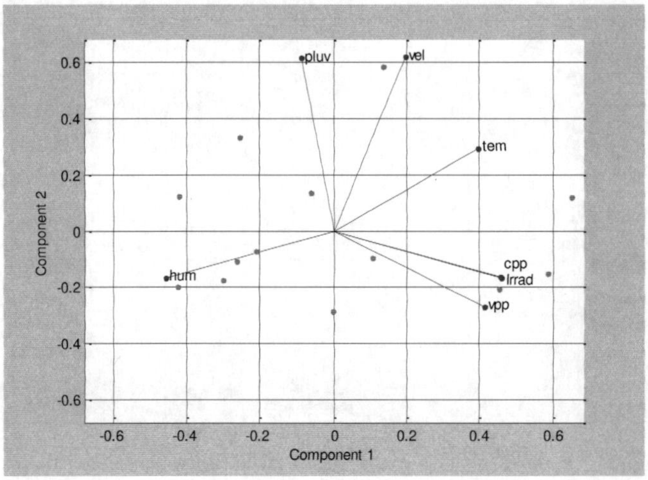

Figure 19. Monocrystalline Panel main components - rainfall

In Figure 19 is seen more clearly the correlation between all the variables with the power output of the panel, the vectors of variables that are found with a small angle with respect to irradiance have a

IOP Conf. Series: Materials Science and Engineering **437** (2018) 012008 doi:10.1088/1757-899X/437/1/012008

positive correlation, while those who have approached 90 degrees angle have a negative correlation, as it is the case of humidity and rainfall.

5.3. C. Analysis of the efficiency

Once made the principal components analysis between panels power and weather variables and after appreciating the relevance that they have, it was made an estimate of the efficiency of the panels according to incident irradiance. Since irradiance is given depending on the area of solar panels (W/m^2), the first thing that must be done to find the relationship of irradiance according to the area of panels used for this study.

The monocrystalline panel area is determined according its dimensions: 0,630m x 0,541m = 0,34083 m^2. As the magnitude of the irradiance measured by the sensor is given for an area of 1 m^2, it is found a relationship k in (4) with the area of the panel:

$$k = \frac{1\text{m}^2}{0{,}34083\ \text{m}^2} = 2{,}934 \quad (4)$$

According to the technical specifications of the Monocrystalline panel its efficiency is 14.67% and the maximum power output is of 50W at STC (standard test conditions). The 50W are given with an irradiance of 1000 W/m^2, therefore, we can check the value of the constant k finding the value of efficiency with maximum values given in (5).

$$\eta = \frac{\text{Pout}*100}{\text{Pin}} = \frac{\text{Pmax}*\text{k}*100}{\text{Pin}} = \frac{50\text{W}*2{,}934*100}{\frac{1000\text{W}}{\text{m}^2}} = \frac{14670}{1000} = 14{,}67\% \quad (5)$$

In the same way the polycrystalline panel area is determined according to its dimensions: 0,550m x 0,680m = 0,374 m^2. As the magnitude of the irradiance measured by the sensor is given for an area of 1 m^2, it is found the relationship k with the area of the panel (6).

$$k2 = \frac{1\text{m}^2}{0{,}374\ \text{m}^2} = 2{,}673 \quad (6)$$

According to technical specifications of the Monocrystalline panel its efficiency is 13.37% and the maximum power output is of 50W at STC (standard test conditions). The 50W are given with an irradiance of 1000 W/m^2, therefore, we can check the value of the constant k finding the value of efficiency with maximum values given in (7).

$$\eta = \frac{\text{Pout}*100}{\text{Pin}} = \frac{\text{Pmax}*\text{k}*100}{\text{Pin}} = \frac{50\text{W}*2{,}673*100}{\frac{1000\text{W}}{\text{m}^2}} = \frac{13365}{1000} = 13{,}365\% \quad (7)$$

Once factors are determined to find the efficiency in monocrystalline and polycrystalline panels, it was related with the irradiance received.

IOP Conf. Series: Materials Science and Engineering **437** (2018) 012008 doi:10.1088/1757-899X/437/1/012008

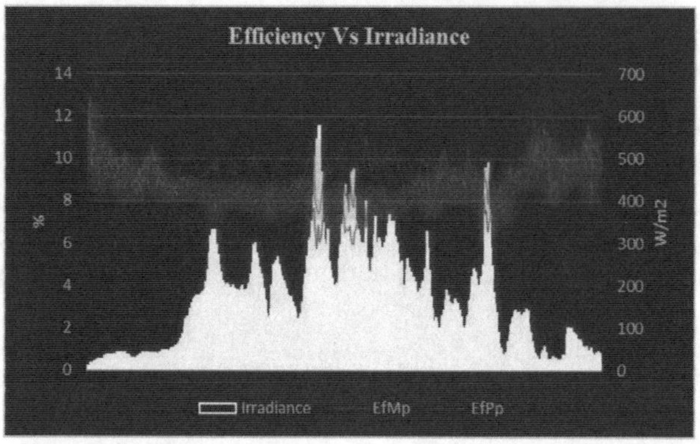

Figure 20. Efficiency Panel according to irradiance

The average efficiency for monocrystalline panel was 9.45% while in the polycrystalline panel was 8.48%. As expected the efficiency of polycrystalline panel it was lower, however, the average efficiency of the two panels was significantly lower than the given by the manufacturer. In the PCA is seen that the panel power has a high positive correlation with the irradiance, while it has a high negative correlation with the humidity and rainfall.

Also, the hours of the day where conditions are more favourable for panel power were determined. In Figure 21 are the vectors of variables and the hours where the irradiance is more predominant. For example, the hours where more power is generated is between 10 a.m. and 11 a.m., while the hours where the humidity was predominant are form 5 a.m. to 6 a.m. and from 4 p.m. to 5 p.m.

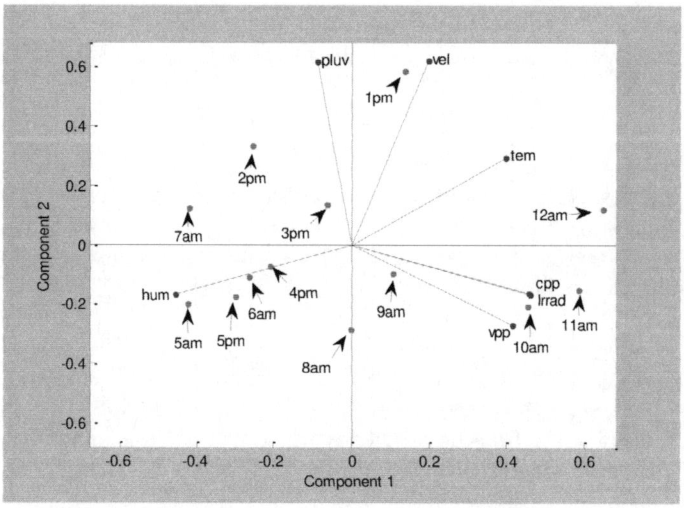

Figure 21. Principal component correlated with efficiency

6. Conclusions

As monocrystalline panel power as polycrystalline panel power, depends not only on their incident irradiance and manufacturing specifications, but is also affected by environmental conditions.
From the study carried out with principal components analysis it was seen the relevance of each climate variable on the generation of power in photovoltaic solar panels of monocrystalline type and polycrystalline type. The variables that had higher positive correlation with the power generated by the

photovoltaic panels were temperature and irradiance, while the variable that had greater negative correlation was the humidity.

Although the rainfall also gave a negative correlation with the generation of power, it was not so critical as the humidity since the rain was not constant. The other variables do not have a noticeable correlation in most of the cases.

Efficiency in actual installation conditions, which differ with the conditions that are defined the specifications of the panels, is significantly reduced with climatic variations such as those in the city of Bogotá. Although, type monocrystalline photovoltaic technology is more efficient, it can be seen that it is around 4% below manufacturer's specification. The same happens with the efficiency of polycrystalline type photovoltaic technology, which reduces in the same proportion.

References

[1] F. Touati, M. Al-Hitmi and H. Bouchech, "Towards understanding the effects of climatic and environmental factors on solar PV performance in arid desert regions (Qatar) for various PV technologies," *Renewable Energies and Vehicular Technology (REVET), 2012 First International Conference on,* pp. 78 - 83, 2012.

[2] W. Mariam and S. Husni, "Influence of Malaysian Climate on the Efficiency of Polycrystalline Solar Cells," *Power and Energy Conference, 2006. PECon '06. IEEE International,* pp. 54 - 57, 2006.

[3] H. Cristian and M. Raducu, "Energy efficiency analysis of various topologies, control techniques and technologies used for photovoltaic panels Part I: On thePV modeling and problem of extracting the maximum power," *Electronics, Computers and Artificial Intelligence (ECAI), 2014 6th International Conference on,* pp. 23 - 28, 2014.

[4] F. Matos and J. Camacho, "A model for semiconductor photovoltaic (PV) solar cells: the physics of the energy conversion, from the solar spectrum to dc electric power," *IEEE Conference Publications,* pp. 352 - 359, 2007.

[5] P. dos Santos and E. Vicente, "Relationship between the shading position and the output power of a photovoltaic panel," *Power Electronics Conference (COBEP), 2011 Brazilian,* pp. 676 - 681, 2011.

[6] F. Eraso Checa, O. Erazo de la Cruz and E. Escobar Rosero, "Study of the relationship between irradiance and power generated by PV panels based on different silicon technologies in the city of Pasto," *Power Electronics and Power Quality Applications (PEPQA), 2015 IEEE Workshop on,* pp. 1 - 6, 2015.

[7] A. S. A. Mark Nixon, "Feature Extraction & Image Processing for Computer Vision," 2012.

[8] J. Heckenbergerova, P. Musilek, J. Marek and J. Rodway, "Principal Component Analysis for Evaluation of Wind Ramp Event Probability," *2014 IEEE Electrical Power and Energy Conference,* pp. 208-212, 2014.

[9] D. Ballabio, "A MATLAB toolbox for Principal Component Analysis and unsupervised exploration of data structure," *Chemometrics and Intelligent Laboratory Systems,* pp. 1-9, 2015.

IOP Conf. Series: Materials Science and Engineering **437** (2018) 012009 doi:10.1088/1757-899X/437/1/012009

Fractal analysis of persistence in fluctuation of levels of the Magdalena River

D A Prada[1], D J Serrano[1], M S Rueda[1], L Duarte[1], A Acevedo[1], S C Prada[2] and J Gómez[3]

[1] Universidad Pontificia Bolivariana, Bucaramanga, Colombia
[2] Universidad Industrial de Santander, Bucaramanga, Colombia
[3] Servicio Nacional de Aprendizaje SENA, Bogotá D.C., Colombia

duwamg.prada@upb.edu.co

Abstract. The measures of water level in a river contributes to understand the dynamic of the discharge in a specific point of the catchment. With this kind of information, it is possible to predict and study future flooding events. The official entity CORMAGDALENA with support of IDEAM report every two days the water level along the Magdalena River. The Hurst coefficient H generates a new statistic methodology, which is based on the tendencies of data series, determining the persistence a dataset and consequently its fractality. This research applies this methodology to a dataset of time, registered by CORMAGDALENA in the municipality of Barrancabermeja, in the Department of Santander. The time has been taken between September 29th of 2011 and March 31st of 2017. According to the value of H we can say if the series of behaviours taken has persistence in the case that H is greater than 0.5, if H is less than 0.5 we will say that the series is anti-persistent. With these values of H it is possible to calculate the fractal dimension associated to the comportment and thus to determine the volatility for the analysis of the risk in terms of the fluctuation of the Magdalena River.

1. Introduction

There are some high complexity systems in the nature, in particular the water basins, because there are some Variables like time, space, weather, geology, vegetation and others are facts linked to their formation process, whose evolution is directly linked with the combination of those forces and weather facts. Due to the dynamic of discharge in specific points of a water basin, the study of water fluctuation levels becomes important, for example in a river, to predict and consider upcoming flood events. The flow of rivers also depends directly of the speed it has, so if the speed is low, the flow will be regular, if the speed increase then it becomes irregular, creating swirls and swirls between swirls. By the Navier-stokes equations the hydrodynamic describes those nonlinear movements initially because the newton laws for fluids. The famous hydrologist Harold Edwin Hurst (1880-1978), studied the fluctuation levels of the Nile River in order to predict fluctuations in order to be prepared for floods and take precautions against drought.

2. Magdalena river

According to the time in the year the level of water in the Magdalena River may change. This kind of phenomena is directly related with periods of drought and floods. Whereby, people can suffer lack of food or damage in the households. The hydrologist Harold Hurst studied the fluctuations of the Nile River with the aim of taking advantage of the reservoir capacities and the measures in drought epochs [1] [2] [3]. For this purpose, Hurst created a statistical methodology which consists of analysing the persistence of a time series [5] according to the duration of its cycles and determining if that time series is fractal [6].

The Magdalena River is one of the most important rivers in Colombia with an average flow is 6700 m^3/s. In fact, the country is considered the fourth in the world in water richness. About 90% of the territory presents rains superiors to 2000 millimetres (mm), with an average evaporation of 1150 mm, and volume of runoff of 2112 km^3 [7]. The river starts in the Colombian Mountains in the department of Huila, and flows into the Caribbean Sea in the Atlantic department. Its length is approximately 1600 km, being 886 navigable. The narrowest part is 2.20 meters, and the widest is 1073 meters [7].

CORMAGDALENA is the entity in charge of the river and is supported by the IDEAM. It resgistrates every two days the level of the Magdalena River, whose measurement stations are in the municipalities of Puerto Berrio, Barrancabermeja, San Pablo, Las Varas, El Banco and Calamar [8]. The data is used to generate time series of depht, with 984 days from September 29, 2011 to March 31, 2017. The fluctuations of the Magdalena river are observed in the municipality of Barrancabermeja, Santander. The analysis is performed to observe the persistence of the time series. Furthermore, it is required to execute a Jarque-Bera test as an examination of normality. Then, it is estimated the fractality in the time series and then the analysis of the rescaled range. After that, it is done the calculation of the Hurst coefficient, and estimated the fractal dimension associated to the fluctuation levels in the Magdalena river, Barrancabermeja.

Study the continuous changes in the fluctuations in the levels of watersheds, it is of great importance because it allows you to adjust predictions about behaviors that can lead to floods or droughts. The first to recognize the characteristics of invariance of the drainage network to establish the laws of scale was Horton, and denoted as: Relationship of fork (R_B), a ratio of length (R_L) and (R_A) [9]. $R_L R_A$Natural objects that have a complex spatial structure can be studied as fractals. Fractal geometry to describe irregular patterns and fragmented, which are repeated at different scales, usually in isotropic [10]. Fractal geometry as a branch of mathematics has been allowed to carry out research in relation to the similarity and dynamic systems, using notions of dimension and orbits in various disciplines such as medicine, engineering, psychology, music, biology, among others. In the particular case of drainage networks two fractal dimensions can be calculated, a fractal dimension (d), which describes the sinuosity of a current and the fractal dimension of the entire network (D), related to the characteristics of branching or compactness of the system [11,12]. With the analysis data between the length of the currents and the area of drainage basins, he found that the Mandelbrot fractal dimension of water courses (d) was equal to 1.2, while the fractal dimension of the entire system (D) was equal or close to 2 [13].

3. Mathematical method

The Convention dimension assigned to certain geometric and physical objects, it is associated with an infinitive variable number, for example, to the cubic number is assigned the triple defined directly by the thickness, width and height thereof and then the size of this object is three. This type of dimension is known as the topological dimension [13, 15].

The fractal dimension, as its name suggests is a fractional dimension and is determinated by a rational number. Long range power law correlations are traditionally measured by a scale parameter or fractal dimension (D). If the time series is self-similar and self-related the parameter d is related to the hurst exponent (H) through the expression D=2-H [16, 17]. Therefore, the hurst ´s exponent is a measured of long range correlation in tile series data and allows distinguishing persistence (correlation), persistence (anti-correlation) or the randomness or data [18]. The original estimate of hurst's exponent was first made in hydrology by Harold Edwin Hurst in 1951 [19], by introducing an empirical relationship called rescale range (R/S). Subsequently, this relationship became the starting point to stablish the classical method R/S (CR/S) developed by Mandelbrot and Wallis in the fractal geometry context [18, 20, 21]. Although CR/S is one of the most popular methods to calculate the Hurst exponent, it has shown some serious limitations for studying the long-range correlation when the time series is not large enough [21, 22].

4. Respect to the test of normality

For the analysis of chaotic or fractal time series, it is required to understand the data applying the linearity and normality test. The Jarque-Bera test is applied for the linearity analysis [23]. This is an asymptotic examination, or large samples based on the ordinary least squares OLS. The idea of persistence of a time series, understood as the tendency to repeat this series (long-term memory), is highly related to the notion of self-sufficiency. Self-splicing is a feature of fractal objects, under which it is possible to observe the irregularity or other properties that maintains the same structure independently of the scale at which it is analysed [4] [13].

The official entity CORMAGDALENA registered 984 data on the fluctuations of the Magdalena River in the municipality of Barrancabermeja between September 29, 2011 and March 31, 2017. The figure 1 shows the graph corresponding to the 984 fluctuations recorded. Likewise, a line represents the average value of the data.

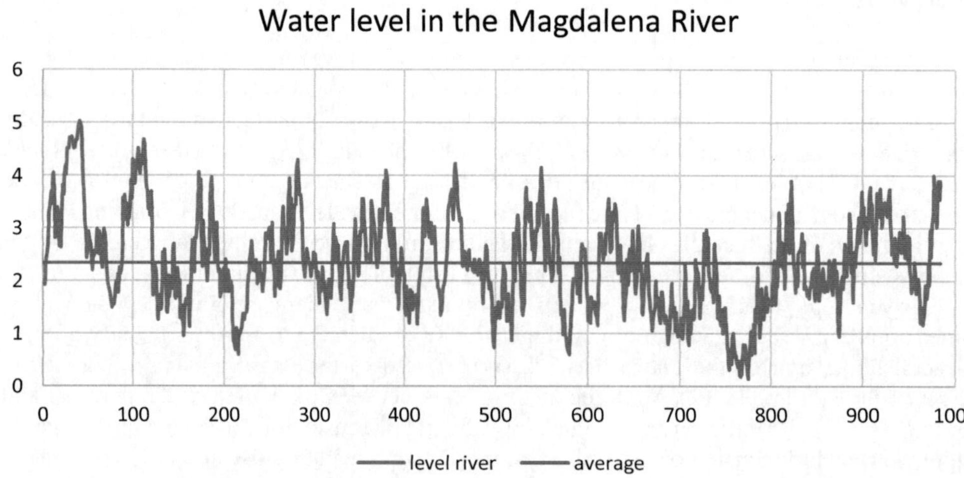

Figure 1: Data on the 984 fluctuations of the Magdalena river in the municipality of Barrancabermeja between September 29, 2011 and March 31, 2017

Thereupon, the Jarque-Bera test is performed to corroborate that the time series is not linear. In other words, it presents fractality. For the data analysed, the asymmetry value was 0.33 and 0.02 for kurtosis. Then the following hypotheses are posed: H0 (null hypothesis): Mean = 0; Variance = 1, which means normal distribution; H1: Mean ≠ 0; Variance ≠ 1, it is not normal distribution. The following equation is used to find the value of the Jarque-Bera test (JB):

$$JB = n\left[\frac{S^2}{6} + \frac{(K-3)^2}{24}\right] \quad (1)$$

Where S represents the asymmetry and K the kurtosis. Therefore, the value of the test is 381.956, as the statistic JB> 6, it is rejected the null hypothesis. Thus, the series does not present any normality.

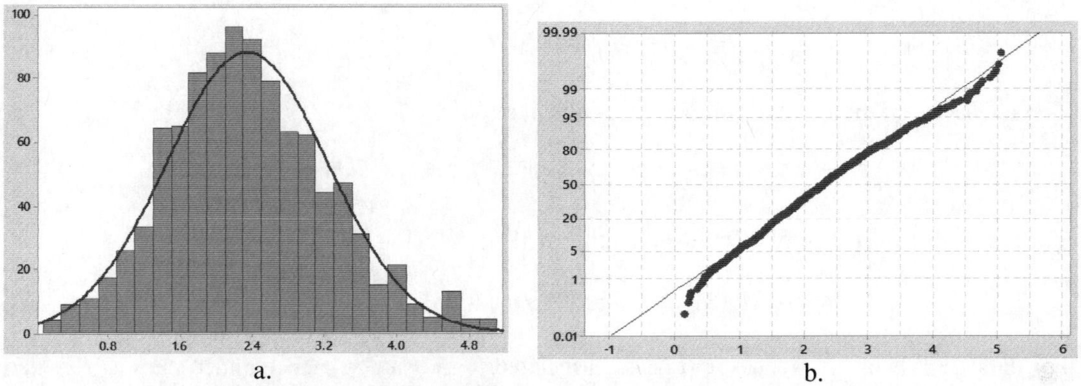

Figure 2. a. Histogram with normal curve of levels of the Magdalena River. b. Graph of probability of levels of the Magdalena River, proof Kolmogorov-Smirnov

Besides the auto-similarity, the fractal objects present another important characteristic, the fractal dimension. The Euclidean or topological dimension is one that assigns to a certain geometric and physical objects, a whole number, for example in a three-dimensional object that has thickness, width and height has three topological dimension. On the other hand, the fractal dimension assigns rational values, due to its irregularity, to certain geometric or physical objects. For example, the measurement of the perimeter of the coast of an island, the traditional measurement method does not allow to use a scale small enough, so that it can be measured in the same way, whereas with the fractal dimension it is possible to assign a value to such presence of irregularity of the contour of the same island. [13]

The Hurst Coefficient is one of the methods which it is possible to calculate the fractal dimension, associated to a time series for statistically self-similar data [24]. Using the accumulated sum and rescaled range method, the coefficient also allows measuring volatility, understand the maximum, and minimum peak values to which data are to be found in a time series with respect to the average value of the same, for the risk analysis of a time series [24]. The persistence of a time series depends on the value of the Hurst coefficient (H). If $0 < H < 0.5$ the Hurst exponent has a high fractal dimension, it will exist for a time series with antipersistent behaviors. A larger exponent of H, $0.5 < H < 1$ has a lower fractal dimension and thus a persistent analysed [24] [5] [25]. According to the statistical mechanics, if H is equal to 0.5, the series presents a random path [6].

The following is CR/S method the time series described in [26] under consideration $X: \{x_i\}$ is composed by N values. The full-time series is divided into windows of size M. The number of windows is defined by $s \equiv N/M$ and therefore there are s windows of data Y_j, with $j = 1,2, \ldots, s$. Defining the vector $k = (j-1)M + 1, (j-1)M + 2, (j-1)M + 3, \ldots, (j-1)M + M$, the average over each window is calculated as:

$$\bar{Y_j} = \frac{1}{M} \sum_k x_k \quad (2)$$

The profile or sequence of partial summations $Z_j: \{z_n\}$, with $n = 1,2 \ldots M$, is defined as the cumulative summation minus the average of the corresponding window

$$z_n = \sum_k^n \{x_k - \bar{y}_j\} \tag{3}$$

The range R_j of the window is defined as the maximum minus the minimum data point of the profile

$$R_j \equiv max\{\,Z_j\} - \min\{\,Z_j\} \tag{4}$$

The standard deviation of each window σ_j is given as

$$\sigma_j = \left[\frac{1}{M}\sum_k (x_k - \bar{y}_j)^2\right]^{\frac{1}{2}} \tag{5}$$

The rescaled range is described by the quantity $(R/S)_M$, which is defined as

$$(\mathrm{R/S})_\mathrm{M} \equiv \mathrm{mean}\ (R_j/\sigma_j) \tag{6}$$

For the case in which a stochastic process associated to the data sequence under study is rescaled over a certain domain $M \in \{M_{min}, M_{max}\}$, the R/S statistics follows the power law

$$(\mathrm{R/S})_\mathrm{M} = \mathrm{a}M^H \tag{7}$$

Herein, is a constant and H is the Hurst exponent which represents a fractal measure of long-range correlations in the analysed released.

5. Results

Due to the series presents fractality, the series of 984 data is partitioned into 4 subgroups [2]. The first subgroup formed by the first 246 data included between the 29th of September 2011 and the 16th of March 2013. With these data the standard deviation was calculated, and the rescaled range. The group 2 consists of the first 492 fluctuations recorded, whose dates vary between September 29, 2011 and July 21, 2014. For the group 3, a number of 738 fluctuations that were recorded between September 29, 2011 and November 25, 2015 and finally, group 4 is formed by the 984 fluctuations recorded during the time between September 29, 2011 and March 31, 2017.

Table 1. Data of the standard deviation, released number and rescaled range

Subgroup	Number of released	Rank rescaled	Max	Min	Average
1	246	78.96	76.86	-2.10	2.66
2	492	90.61	84.95	-5.65	2.60
3	738	108.84	107.82	-1.02	2.41
4	984	169.21	128.71	-40.49	2.34

Table 2 shows the data needed for the calculation of the coefficient of hurts for this time-series, as well as compare the natural logarithm of the data of each of the subgroups generated versus the natural logarithm of the ratio between the rescaled range and standard deviation.

IOP Conf. Series: Materials Science and Engineering **437** (2018) 012009 doi:10.1088/1757-899X/437/1/012009

Table 2. Related data from the four groups depending on the range and Standard deviation. Calculation of the natural logarithm of the data number and the natural logarithm of the quotient between the rescaled range and the standard deviation

Subgroup	Number data	Rescaled range	Standard Deviation	Ln(Num)	Ln(R/S)
1	246	78.96	1.04	6.27	5.43
2	492	90.61	0.87	6.96	6.16
3	738	108.84	0.87	7.37	6.52
4	984	169.21	0.89	7.66	6.81

With this data we make a new graph in which we compare the natural logarithm of the data of each of the subgroups generated versus the natural logarithm of the ratio between the rescaled range and standard deviation, these data are related in Figure 3.

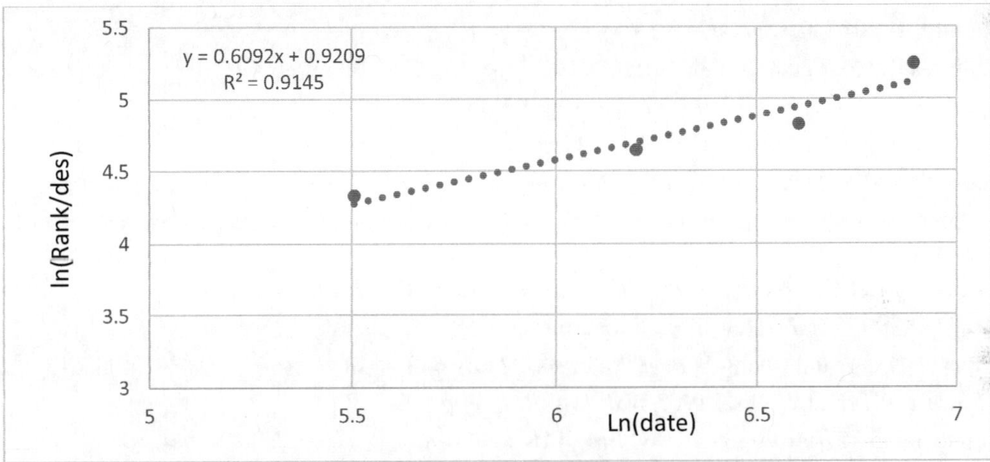

Figure 3. Logarithmic linear regression of the data supplied in Table 2

Performing the linear adjustment, with the regression, the value of the Hurst coefficient is obtained as the slope of the resulting line.

6. Conclusions

Along the period studied, the 984 fluctuations recorded by the official entity CORMAGDALENA between September 29, 2011 and March 31, 2017, was verified that this series did not present linearity by the Jarque-Bera test through equation (1). This implies that the time series could be a fractal. Due to the fact, there is a certain degree of self-sufficiency. In other words, the fractal dimension can be assigned to this time series. The value of the Hurst coefficient, which was calculated using the accumulated sums, rescaling range and logarithmic linear regression, is equal to 0.6092, $H > 0.5$, which means, the time series is persistent or has long-term memory. Given the condition of the equation $D + H = 2$, where D is the fractal dimension and H is the Hurst coefficient, we have that the fractal dimension of the series is $D = 1.3908$. From the data above presented, it can be concluded that the series is persistent under a volatility of 69.54%, which implies an average risk in future behaviour.

IOP Conf. Series: Materials Science and Engineering **437** (2018) 012009 doi:10.1088/1757-899X/437/1/012009

References

[1] Braun E 2011 *Caos, fractales y cosas raras* (Mexico: Fondo de Cultura Económica) p 92

[2] Rodriguez E 2012 *Lat. Amer. Not. Adm.* **3** pp 42-50

[3] Hurst H 1951 Trans. Am. Soc. Civ. Eng. 116 pp 770-808

[4] Mandelbrot B 2009 *La Geometría Fractal de la Naturaleza* **2** (España: Metatemas)

[5] Peters E 1991 *Chaos and Order in the Capital Markets* (EUA: John Wiley & Sons, Inc.)

[6] Peters E 1994 *Fractal market analysis: applying chaos theory to investment and economics,* (New York: Willey)

[7] E. B. Duffo, «Banco de la República, actividad cultural,» 1 Junio 2013. [En línea]. Available: http://www.banrepcultural.org/blaavirtual/revistas/credencial/junio-2013/el-rio-magdalena. [Último acceso: 1 mayo 2017].

[8] Corporación Autónoma del Río Grande de la Magdalena, Presidencia de la República, «CORMAGDALENA,» 5 mayo 2016. [En línea]. Available: http://niveles.cormagdalena.gov.co/. [Último acceso: 2 abril 2017].

[9] Ramirez R 2017 *Min. and Geo.* **33** pp 279-292

[10] Ramirez R, Rodriguez A and Ordaz A 2017 *Min. and Geo.* **33** pp 36-49

[11] Barbera P and Rosso R 1989 *Wat. Res. Res.* **25** pp 735-741

[12] Tarboton D 1996 *J. Hyd.* **187** pp 105-117

[13] Mandelbrot B 2006 *Los objetos fractales; forma, azar y dimensión* 6 (España: Metatemas)

[14] Falconer K 2006 *Fractal geometry. Mathematical foundations and applications, second edition* (India: Wiley)

[15] Barnsley M 2012 *Fractals Everywhere* (United States: Dover Publications)

[16] Erdf J 1988 *Fractals* (New York: Plenum Press)

[17] Chen C, Lee Y and Chang Y 2008 *Phys. A* **387** pp 4643-4648

[18] Mandelbrot B and Wallis J 1968 *Wat. Res. Res.* **4** pp 909-918

[19] Hurst H 1951 *Trans. Am. Soc. Civ. Eng.* **116** pp 770-808

[20] Mandelbrot B and Wallis J 1969 *Wat. Res. Res.***5** pp 321-340

[21] Mandelbrot B and Wallis J 1969 *Wat. Res. Res.* **5** pp 967-988

[22] Sanchez M, Trinidad J and García J 2008 Phys. A 387 pp 5543 -5551

[23] Gujaratí D, Porter D 2010 *Econometría* **5** (México: McGraw Hill)

[24] Mandelbrot B, Hudson R 2010 *Fractales y finanzas, una aproximación matemática a los mercados: arriesgar, perder y ganar* (España: Metatemas)

[25] Mandelbrot B 1997 *Fractals and Scalind in Finance* (EUA: Springer)

[26] Martin L, Aranda N and Quimbaya C 2015 *Phys. A* **421** pp 124 -133

IOP Conf. Series: Materials Science and Engineering **437** (2018) 012010 doi:10.1088/1757-899X/437/1/012010

Modeling and optimization of essential amino acids effects on productive performance in second-cycle hens

W Morales [1], I C Ospina[2], J J Méndez[1] and H A Váquiro[1]

[1] Universidad del Tolima, Barrio Santa Helena Parte Alta, A.A. 546, Ibagué, CP 730006299, Colombia
[2] Universidade Estadual de Maringá, Av. Colombo, 5790, Maringá, PR 87020-900, Brazil

havaquiro@ut.edu.co

Abstract. In this study, a model based on a modification of the Gompertz equation was used to estimate the maximum percentage of egg production and digestible requirements of lysine (Lys), methionine + cystine (Met+Cys) and threonine (Thr) of H&N brown laying hens during 16 weeks of production. The productive performance model was adjusted to field data of 23 treatments, which were defined from a composite central design that combines five levels of Lys, Met+Cys, and Thr ranged from 0.636% up to 1.000%; 0.579% up to 0.910%; 0.483% up to 0.760%, respectively. The proposed model showed to be adequate to evaluate the efficiency in the use of digestible amino acids for egg production and can be a useful tool to diagnose the effectiveness of the nutritional requirements in each of the production phases.

1. Introduction

The poultry industry has shown significant growth in recent years, although production costs are high and most of them are based on feed costs, which can exceed 70% of variable costs [1]. One of the strategies to reduce such production costs is forced molting, which is an important management tool that offers economic benefits associated with additional egg laying cycles and provides the producer with flexibility in handling to respond with rapid changes in the egg market, as well as changes in feed and input costs [2].

Limiting amino acid requirements in diets of second-cycle hens differ from those of first-cycle birds because of their special conditions, including differences in weight gain, egg production and egg size. Based on the ideal protein concept, it is claimed that there is a relationship between lysine and other amino acids and this relationship should be considered to avoid reduced bird performance [3]. However, the interactions between amino acids can be modified based on factors such as age, sex, and environment [4]. Methionine is the first limiting amino acid in poultry diets, followed by lysine. Methionine is required for the biosynthesis of important substances involved in growth [5]. Dietary deficiency of methionine reduces egg production and body weight and increases fat deposition in the liver of laying hens. However, factors such as genetic line, environmental temperature and energy content, as well as the presence of anti-nutritional factors and the processing of raw materials used in the diet, influence the requirement of methionine + cystine in laying hens [6]. Threonine plays an important role in gastrointestinal health and its optimization offers greater protection to bacterial infections of the intestine. The high requirement of gut for threonine is expected because this amino acid accounts for 30% of the total amino acid content of mucin [7,8].

Researches regarding amino acid requirements of laying hens take into account a complexity of factors that can act collectively or individually. Optimal levels of amino acids may vary depending on genetic factors, immune stress, age, dietary and climatic factors. In addition to these factors, estimates of amino acid requirements may also depend on the mathematical model used in their prediction, as some models tend to overestimate requirements and others to underestimate [9]. Digestible amino acids requirements of second-cycle hens have been estimated for the genetic lines Lohman Brown and Shaver Brown in 0.681% and 0.790% of lysine [10,11], 0.698% and 0.588% of methionine + cystine [10,12], 0.460% and 0.525% threonine [12,13], respectively. However, nutritional requirements of second-cycle hens during production phase are unknown for other genetic lines under specific production conditions and it is necessary to determine amino acid requirements to optimize the physiological status of birds, regeneration of their tissues after molting, and increased production. Currently, dose-response is the methodology most used to establish the optimal levels of amino acids in experimental trials. This method consists of the gradual increase of the amino acid to the basal diet deficient in the amino acid to be studied [14]. However, other authors [15] suggest that the dose-response approach ignores the effects of interaction between amino acids and recommend methodologies evaluating multiple nutrients in the same assay.

Mathematical models are strategies that allow the evaluation of egg production during the laying cycle as well as the prediction of financial losses caused by a fall in productive performance [16]. The Gompertz model has been one of the most used equations in the poultry industry to represent mathematically biological and economic phenomena associated with growth, due to the biological significance of its parameters, the relative ease with which the model is adjusted and the robustness in describing the trend of experimental or field data [17]. In this study, a procedure was developed based on the modification of the Gompertz model to determine the percentage of egg production in response to essential amino acids supplementation. The procedure was applied to determine the requirements of digestible lysine, methionine + cystine and threonine levels for egg production of H&N brown laying hens during the second production cycle.

2. Materials and methods

2.1. Field work and experimental information

The field work was carried out at the San Carlos Poultry Farm (San Pedro, Antioquia) located at 2475 above sea level with a temperature of 14 °C and relative humidity of 79% annual average. A total of 1380 92-week-old H&N brown laying hens during the second production cycle were housed in cages with commercial production conditions. The number of eggs produced was recorded weekly for 16 weeks. The percentage of egg production (P) was calculated weekly from the relation between the number of eggs and the number of birds per experimental unit.

2.2. Experimental design

Dietary digestible essential amino acids concentrations (three factors: lysine, methionine + cystine and threonine) were combined using a composite central design. A design matrix was obtained with 23 treatments (Table 1) combining five levels of lysine (*Lys*), methionine + cystine (*Met+Cys*) and threonine (*Thr*) ranged from 0.636% up to 1.000%; 0.579% up to 0.910%; 0.483% up to 0.760%, respectively. Five replicates were performed per treatment in experimental units of 12 birds.

2.3. Formulation of diets

The diets corresponding to each of the 23 treatments were formulated with corn and soybean cake, corn gluten and palm oil according to the nutritional recommendations of the birds as proposed by Rostagno et al. [18]. The essential amino acids (*Lys*, *Met+Cys*, *Thr*) were fixed at their respective five levels by the addition of commercial products containing supplemental amino acids. The remaining seven essential amino acids were fixed by changing the level of kaolin in the diet.

2.4. Modeling and Statistical Analysis
The mathematical modeling of the egg production (Figure 1) as a function of time was performed for each of the treatments using a Gompertz model with modifications (Equation 1 and 2).

$$P = P_m \exp\left\{-\exp\left[\frac{\mu e}{P_m}(\lambda - t) + 1\right]\right\} \tag{1}$$

$$P_m = \gamma_1 + \gamma_2 t \tag{2}$$

In equations (1) and (2), P is the egg production (%), t is the time (weeks), λ is the initial lag time (weeks), μ is the maximum growth rate (% week^{-1}), e is the base of the natural logarithm, P_m is a function representing the behavior of egg production during the maturity phase (%), γ_1 (%) and γ_2 (% week^{-1}) are parameters of P_m function.

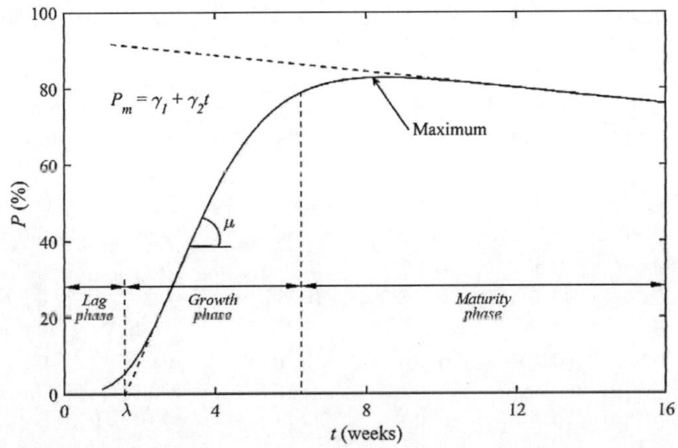

Figure 1. Production curve parameters for the modified model of Gompertz

The model parameters (λ, μ, γ_1 and γ_2) were estimated using "nlinfit" function of the Statistic Toolbox of MATLAB® R2016b (The MathWorks Inc., Natick, MA, USA) for each treatment. The dependence of model parameters with the amino acid concentrations was assessed to obtain a unique model in terms of time and *Lys*, *Met+Cys* and *Thr*. The 95% confidence intervals of the estimated parameters were determined using the "nlparci" function and the "lillietest" function was used to determine if the residuals followed a normal distribution. The adjusted coefficient of determination (Equation 3) and the mean relative error (Equation 4) were used to evaluate the goodness of fit and accuracy of the estimation, respectively.

$$R_{adj}^2 = 1 - \frac{S_{yx}^2}{S_y^2} \tag{3}$$

$$MRE = \frac{100}{n}\sum_{i=1}^{n}\frac{\left|P_i - \hat{P}_i\right|}{P_i} \tag{4}$$

In equations (3) and (4), S_y is the standard deviation for the sample, S_{yx} is the standard deviation for the model estimation, $\hat{P}$ is the egg production estimated by the model (%), and n is the number of experimental data points used in the regression analysis.

3. Results and discussion

The estimated parameters and goodness-of-fit statistics of the egg-production model are shown in Table 1. The model was suitable for represent the egg production for the whole performance curve in all the treatments ($R^2_{adj} > 0.86$, $MRE < 10\%$).

Table 1. Estimated parameters and goodness-of-fit statistics of the egg-production model

Treatment	*Lys* (%)	*Met+Cys* (%)	*Thr* (%)	γ_1 (%)	γ_2 (% week^{-1})	μ (% week^{-1})	λ (weeks)	R^2_{adj}	*MRE* (%)
1	0.636	0.579	0.483	64.85	0.926	11.32	0.204	0.942	8.29
2	0.636	0.579	0.760	78.54	0.215	13.27	0.759	0.960	5.66
3	0.636	0.744	0.622	81.07	-0.220	16.23	0.000	0.904	6.46
4	0.636	0.910	0.483	79.74	-0.065	15.43	0.001	0.902	6.71
5	0.636	0.910	0.760	86.88	-0.416	24.11	1.700	0.965	11.3
6	0.710	0.646	0.539	98.97	-1.281	13.16	1.293	0.945	9.09
7	0.710	0.646	0.704	83.08	-0.123	22.42	1.786	0.967	12.2
8	0.710	0.843	0.539	96.27	-0.863	14.45	1.196	0.948	8.20
9	0.710	0.843	0.704	75.56	0.258	15.26	0.000	0.928	5.04
10	0.818	0.579	0.622	72.03	0.423	16.94	0.006	0.918	6.09
11	0.818	0.744	0.483	83.35	-0.359	15.35	0.000	0.901	7.87
12	0.818	0.744	0.622	89.48	-0.461	15.81	0.000	0.902	8.08
13	0.818	0.744	0.760	79.61	-0.087	22.91	0.006	0.920	4.76
14	0.818	0.910	0.622	86.52	-0.454	14.40	0.000	0.863	9.04
15	0.926	0.646	0.539	71.64	0.191	16.17	0.069	0.907	6.77
16	0.926	0.646	0.704	66.25	0.945	16.37	0.076	0.907	7.25
17	0.926	0.843	0.539	87.75	-0.376	13.36	0.000	0.927	7.09
18	0.926	0.843	0.704	100.81	-1.563	12.34	0.000	0.939	7.72
19	1.000	0.579	0.483	83.22	-0.287	13.69	0.696	0.938	8.00
20	1.000	0.579	0.760	81.99	-0.118	12.91	0.494	0.955	8.21
21	1.000	0.744	0.622	88.76	-0.638	13.69	0.566	0.971	6.21
22	1.000	0.910	0.483	91.53	-0.433	11.38	0.477	0.955	9.26
23	1.000	0.910	0.760	68.44	0.887	14.63	0.925	0.960	9.26

Each estimated parameter was submitted to a stepwise regression to write it in terms of *Lys*, *Met+Cys* and *Thr* contents. These expressions were used to fit simultaneously the experimental data of all treatments and obtain a general model corresponding to equations (5) to (8). In this generalized egg-production model are only considered statistically significant terms. The estimated parameters and 95% confidence intervals of the generalized egg-production model are shown in Table 2.

$$P = P_m \exp\left[-\exp\left(1 - \frac{\mu e}{P_m} t\right)\right] \quad (5)$$

$$\mu = \beta_1 + \beta_2 Lys^2 \quad (6)$$

$$\gamma_1 = \beta_3 + \beta_4 Lys + \beta_5 (Met+Cys) + \beta_6 Lys(Met+Cys) \quad (7)$$

$$\gamma_2 = \beta_7 + \beta_8 Lys + \beta_9 (Met+Cys) + \beta_{10} Lys(Met+Cys) \quad (8)$$

Table 2. Estimated parameters and confidence intervals of the generalized egg-production model

Parameter	Estimated value	95% Confidence interval
β_1	17.704	15.909, 19.499
β_2	-5.7908	-8.0382, -3.5433
β_3	-133.46	-251.86, -15.054
β_4	234.85	83.342, 386.36
β_5	245.08	85.979, 404.19
β_6	-260.77	-463.04, -58.496
β_7	15.466	6.2847, 24.648
β_8	-17.474	-29.088, -5.8593
β_9	-17.980	-30.285, -5.6761
β_{10}	19.802	4.3157, 35.287

The lag time (λ) did not show statistical significance in the model. On the other hand, the production rate in the growth phase (μ) was influenced only by *Lys*, while the behavior of the maturity phase was dependent on *Lys*, *Met+Cys* and their interactions.

The generalized egg-production model exhibited a satisfactory fit of the experimental data with high adjusted coefficient of determination (R^2_{adj} = 0.885), as well as high accuracy of the estimation (*MRE* = 5.4%). According to the residual analysis, 84.8% of the results were between -7.72% and 10.2%. As it can be seen in Figure 2, the residuals appeared scattered randomly around zero with absolute values less than 21.2%. Thus, corroborate the suitability of the proposed model to accurately predict the egg-production in terms of digestible amino acid requirements.

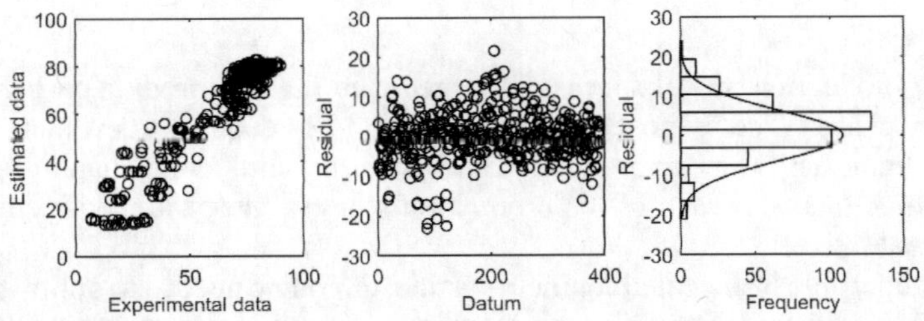

Figure 2. Residual analysis for the egg-production model

The simulated results for the generalized egg-production model (Figure 3) showed the importance of each of the essential amino acids evaluated for the different stages of the model. In the lag phase, the levels of *Lys* (0.636%), *Met+Cys* (0.800%) and *Thr* (0.483%) are adequate to supply the needs of the commercial laying hen. In the growth phase, production depends only on *Lys* for the model and the levels of *Lys* (0.636%), *Met+Cys* (0.900%) and *Thr* (0.483%) are adequate to achieve a production above 76%. In the maturity phase, production depends on *Met+Cys*, and *Lys* levels and its interactions. Levels of *Met+Cys* (0.646%), *Lys* (0.636%) and *Thr* (0.483%) are suitable for a production of at least 80%.

The estimated optimal level of *Lys* for egg production for 16 weeks was lower to those recommended by Schneider [11] of 0.79% for semi-heavy laying hens. Likewise, Schmidt et al. [10] found quadratic effect of *Lys* levels on egg production and estimated the optimal level at 0.681% for semi-heavy laying hens. This value was similar to the level recommended in this study of 0.636% digestible *Lys* for H&N semi-heavy laying hens. However, Cupertino et al. [19] estimated lower values for digestible Lys of 0.692% on egg production.

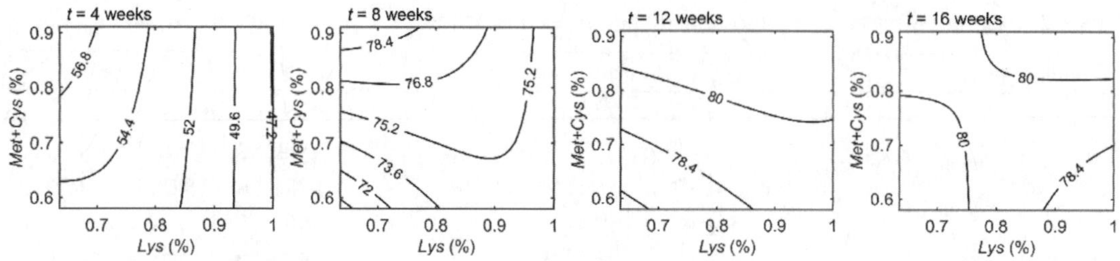

Figure 3. Effect of lysine (*Lys*) and methionine + cystine (*Met+Cys*) on egg-production

There were estimated digestible *Met+Cys* levels of 0.646% for maximizing the egg production of H&N brown hens during 16 weeks of production. This estimated level was different to estimated values reported by Schmidt et al. [20] and Polesse et al. [12]. These authors recommended values of 0.698% and 0.580% digestible *Met+Cys* for semi-heavy laying hens to optimize egg production, respectively.

Dietary digestible *Thr* level of H&N semi-heavy laying hens was estimated at 0.483%. Schmidt et al. [20], who evaluated *Thr* levels ranged from 0.380% up to 0.512% in semi-heavy laying hen diets, estimated a lower digestible *Thr* level of 0.467%. However, Agustini et al. [13] recommended a higher *Thr* level of 0.501% for egg production of semi-heavy laying hens.

Although the first limiting amino acids in laying hen diets have been long studied, there are discrepancies in the estimated *Lys*, *Met+Cys* and *Thr* requirements in second-cycle hen diets from different studies, mainly for dietary *Met+Cys* and *Thr* levels. The interactions effects between these amino acids may be an important factor of variation in these estimates as corroborated in this study. Therefore, the study of *Lys*, *Met+Cys* and *Thr* requirements at the same time is an important point to be considered for the establishment of appropriate levels of these amino acids in the diet.

4. Conclusions

The generalized egg-production model successfully predicted the egg production behavior in the lag, growth, and maturity phases. The proposed model showed to be adequate to evaluate the efficiency in the use of digestible amino acids for egg production and can be a useful tool to diagnose the effectiveness of the nutritional requirements in each of the production phases. Also, the model could be applied in other species of animal interest.

Dietary levels of digestible lysine, methionine + cystine, threonine needed to optimize egg production for 16 weeks of H&N semi-heavy laying hens were estimated at 0.636, 0.646 and 0.483 % respectively.

5. Acknowledgments

The work was sponsored by the poultry farm “San Carlos” and the university of Tolima.

6. References

[1] FENAVI 2013 Huevo: inversión en crecimiento *Rev. Avic.* **203** 15–8

[2] Park S Y, Kim W K, Birkhold S G, Kubena L F, Nisbet D J and Ricke S C 2004 Induced moulting issues and alternative dietary strategies for the egg industry in the United States *World Poultry Sci. J.* **60** 196

[3] Pastore S M 2014 Níveis de lisina digestível em rações para galinhas poedeiras leves no período de 24 a 40 e de 42 a 58 semanas de idade (Minas Gerais, Brazil: Universidade Federal de Viçosa)

[4] Tejedor A A 2002 Exigências nutricionais de metionina+ cistina, de treonina e de arginina para frangos de corte nas diferentes fases de criação (Minas Gerais, Brazil: Universidade Federal de Viçosa)

[5] Baker D H, Fernández S R, Webel D M and Parsons C M 1996 Sulfur amino acid requirement and cystine replacement value of broiler chicks during the period three to six weeks

posthatching *Poultry Sci* **75** 737–42

[6] Filho J J, da Silva J H V, da Silva E L, Luis M, Ribeiro G, Martins T D D and Rabello C B V 2006 Exigências nutricionais de metionina+ cistina para poedeiras semipesadas do início de produção até o pico de postura *Rev. Bras. Zootecn.* **35** 1085-90

[7] Good L 2013 The effects of actigen® and threonine supplementation on growth parameters, immune function, and intestinal health in monogastrics (Lexington: University of Kentucky)

[8] Faure M, Moënnoz D, Montigon F, Fay LB, Breuillé D, Finot PA, Ballèvre O and Boza J 2002 Development of a rapid and convenient method to purify mucins and determine their in vivo synthesis rate in rats *Anal. Biochem.* **307** 244–51

[9] Venturini K S 2011 Modelos para estimar as exigências de lisina digestível para poedeiras comerciais (São Paulo, Brasil: Universidade Estadual Paulista "Julio De Mesquita Filho")

[10] Schmidt M, Gomes P C, Rostagno H S, Albino L F T, Nunes R V and Cupertino ES 2009 Nutritional requirement of digestible lysine for brown-egg laying hens in the 2nd production cycle. *Rev. Bras. Zootecn.* **38** 1956–61

[11] Schneider S E 2011 Exigência de lisina digestível para poedeiras semipesadas (Paraná, Brazil: Universidade Estadual do Oeste do Paraná)

[12] Polese C, Nunes R V, Vilela C G, Murakami A E, Agustini M A, Takahashi S E, Vilela V O, Souza C and Schneider S E 2012 Quantidade de metionina+ cistina digestível para poedeiras semipesadas de 75 a 91 semanas de idade *Arq. Bras. Med. Vet. Zootecn.* **24** 1682–90

[13] Agustini M A B, Nunes R V, Vilela C G, Takahashi S E, Vilela V O, Santos Bueno R and Polose C 2014 Nutritional levels of digestible threonine in brown-egg laying hens from 75 to 90 weeks of age *Semina Ciências Agrárias* **35** 3449–56

[14] Bendezu H C P 2012 Modelos para estimar a ingestão ótima de metionina+ cistina para poedeiras comerciais (São Paulo, Brasil: Universidade Estadual Paulista "Julio De Mesquita Filho")

[15] De Leon A C, Kidd M T and Corzo A 2010 Box-Behnken Design: alternative multivariate design in broiler nutrition research *World Poultry Sci. J.* **66** 699–706

[16] Savegnago R P, Nunes B N, Caetano S L, Ferraudo A S, Schmidt G S, Ledur M C and Munari D P 2011 Comparison of logistic and neural network models to fit to the egg production curve of White Leghorn hens *Poultry Sci.* **90** 705–11

[17] Silva E P, Sakomura N K, Dorigam J C P, Malheiros E B, Fernandes J B K and Araujo J A 2014 A procedure to evaluate the efficiency of utilization of dietary amino acid for poultry *Acta Sci. Anim. Sci.* **36** 163–9

[18] Rostagno H S, Teixeira L, Lopes Donzele J, Gomes P C, Oliveira R F, Lopes D C, Ferreira A S, Toledo S L and Euclides R F 2011 Tablas brasileñas para aves y cerdos: composición de alimentos y requerimientos nutricionales 3rd ed (Minas Gerais, Brazil: Universidad Federal de Viçosa)

[19] Cupertino E S, Gomes P C, Albino L F T, Donzele J L, Mello H H C, Schmidt M and Calderano A A 2009 Exigência nutricional de lisina digestível para galinhas poedeiras de 54 a 70 semanas de idade *Rev. Bras. Zootecn.* **38** 480-7

[20] Schmidt M, Gomes P C, Rostagno H S, Albino L F T, Nunes R V U and Brumano G 2009 Níveis nutricionais de metionina+cistina digestível para poedeiras semipesadas no segundo ciclo de produção *Rev. Bras. Zootecn.* **38** 162-8

IOP Conf. Series: Materials Science and Engineering **437** (2018) 012011 doi:10.1088/1757-899X/437/1/012011

Changes on chemical composition of cocoa beans due to combined convection and infrared radiation on a rotary dryer

J M Castellanos[1], C S Quintero[1] and R Carreno[1]

[1]Universidad Pontificia Bolivariana, Bucaramanga, Colombia

javier.castellanos@upb.edu.co

Abstract. Cocoa production around the world is characterized by a lack of technical facilities for post-harvest activities, and therefore a good quality of beans is hard to achieve. This situation can be changed by using technological solutions, particularly when they are adapted to specific needs of cocoa producers. Mechanical dryers used in intensive cocoa production exceed the smallholders′ requirements and because of the large amounts of energy they consume, there is a substantial economic and environmental impact. At this regard, improvement of alternative mechanical drying techniques requires a careful assessment of their impact on cocoa quality. Then this work looks forward to evaluating the chemical changes on cocoa beans due to the combined use of convection and radiation in a rotary dryer. It highlights not only the experimental set ups but also the analysis methods involved in the chemical characterization. High Performance Liquid Chromatography (HPLC) and Spectrophotometry are the chemical analysis tools used in this study. On one hand, total polyphenols content is determined by means of a spectrophotometric method based on the use of the Folin & Ciocalteu's reagent and a mix of hexane-isopropanol at pH=9 as extraction solution. Similarly, in order to identify the impact of the combined drying strategy on the concentration not only of Polyphenols (Catechin and Epicathechin) but also of Methylxanthines (Caffeine and Theobromine) a chemical identification and quantification of these compounds are done using HPLC with UV detector and a C-18 column, using water, methanol and acetic acid as mobile phase.

1. Introduction

According to its industrial process, the supply chain of cocoa is divided in three phases: Upstream, Middlestream and Downstream [1]. The upstream is related primarily to the farmers work, and it begins after harvest, when cocoa beans are extracted from the pod, to undergo a postharvest process. Then, they are commercialized for industrial transformation purposes that depend on market needs. The postharvest process is composed by several stages. First occurs the fermentation step, where beans are stored in a ventilated place for five days to allow bacteria and yeast to grow on the grain. Well fermented cocoa is featured by a swollen, dark brown, easily breakable bean releasing a pleasant aroma and offering a bitter taste [2]. Afterwards, the drying process in performed to reduce water activity in the bean, and finally, there is a selection step to eliminate impurities such as soil particles and moldy or broken beans, among others. Drying step plays a major role to assure quality of the final product and sun drying is the most common method used by small producers. When the weather is in good conditions, solar drying may be efficient and cheap, but there is a lack of ability to control the process, and a possible degradation due to biochemical and microbiological reactions and to insect infestation. It is important to point out that if production has to be augmented, there would be a serious limitation due to an increase of labor cost and area requirements. Additionally, lack of control results in different moisture content among beans which is a major problem due to the fact that traders mix beans because they store the row material for some time until its amount is economically transportable to bigger traders' warehouses.

Cocoa quality depends on every one of the production processes, from the farming where is cultivated until the transportation to industrial facility where is transformed. It is well known the influence of different varieties, the country of origin and farming activities in the characteristics of the cocoa beans, and on the flavor profile of the cocoa derived products [3]. Likewise, many studies have showed that amino acid, sugar, polyphenols of fat contents are dependent of the origin of cacao beans [4–7]. Nevertheless, according to literature, postharvest processing has a strong influence in the final quality of the row material. Drying of cocoa beans is a process of heating which reduces its moisture content to less than 7.5%. Considering the results of the studies related with the variables of the drying process and their effects on cocoa beans and chocolate, there are three principal issues that had been addressed: Duration, Temperature and Method of drying. It is a common practice that the duration of the drying process ends up when the desired moisture content in reached. Unfortunately, the moisture content and the consequential overall quality of the dried cocoa beans might vary considerable between samples [8]. Drying is also a continuation of the oxidative stage of the fermentation process and therefore plays an important role in reducing astringency, bitterness and acidity. During this process, the characteristic brown color of the chocolate is developed due to the enzymatic oxidation of polyphenols [9,10]. The drying rate is of crucial importance for the cocoa beans´ finally quality. When drying rate is too fast, the beans would tend to retain an excessive amount of acids, including acetic acid, which is deleterious of the flavor. By contrast, a slower drying rate would result in low acidity, poorer color and high presence of moulds [9,11–16]. Recent studies identified 58 volatile compounds in cocoa beans at different times and drying temperatures in order to improve the conditions in the processing of cocoa beans. It was determined that six days of fermentation were enough to produce volatile compounds with flavor notes and that drying at 70°C and 80°C after six fermentation days presented a volatile profile similar to the one obtained by sun drying [15].

In reference to the method of drying, many studies that compare natural and artificial drying methods conclude that natural sun drying gives the best result [16,17]. Likewise, the moisture profile during drying and different temperatures and different conditions is available in the literature [11], [13,18,19]. Thus, knowing the conditions of drying it would be possible to predict in real time the ideal duration of drying to reach a standardized moisture content by the use of prediction models like those proposed by several authors [18,20,21]. Artificial drying methods are believed to be able to improve cocoa drying processes but more research have to be done in this area [22]. Most of studies in the literature deals with convective hot air drying. The use of combined convective and IR drying for cocoa beans has not been studied yet but considering some advantages of IR drying techniques, there is an important potential to improve the cocoa drying process. IR radiation could reduce the drying time and it can be applied directly on the surface of the product without heating surroundings.

From a strict chemical point of view, recent works detail different methodologies and equipment to analytically determine specific chemical compounds: Fatty acids, sugars, polyphenols, methylxanthines and aromatic compounds. Volatile compounds are extracted and quantified using combined technique like solid phase micro-extraction (SPME-HS), gas chromatography and mass spectrometry (GC-MS) [15,23]. High performance liquid chromatography (HPLC) and Ultra Performance Liquid Chromatography (UPLC) had been used in various studies to analyze most of the polyphenols present in cocoa and chocolate extracts [5,24–26]. No volatile acids, methylxanthines and carbohydrates had been determined using also HPLC, following well documented procedures [23,26–28]. High resolution gas chromatography -mass spectrometry (HRGC-MS) has been used for identification and quantification of volatile compounds [29]. Recently, methylxanthines, organic acids, amino acids, carbohydrates and polyphenols have been identified using Nuclear Magnetic Resonance (NMR) spectroscopy (1H NMR) [30,31].

Then this work is a preliminary experimental study looking forward to evaluating the chemical changes on cocoa beans due to the combined use of convection and radiation in a mechanical dryer when cocoa beans surface temperature is controlled. It highlights not only the experimental set up but also the analysis methods involved in the chemical characterization. On one hand, total polyphenols content (TPC) are determined using a spectrophotometric method based on the Folin & Ciocalteu's reagent [32]. Likewise, to assess the impact of the combined drying strategy on the concentration not only of Polyphenols ((+)-catechin and (-)-epicathechin) but also of Methilxanthines (Caffeine and Theobromine) a chemical identification and quantification of these compounds are done using High Performance Liquid Chromatography (HPLC), chemical standards and liquid micro-extraction technique. HPLC has been employed in several cocoa studies to successfully identify specific compounds based on their retention time [24].

2. Materials and methods

2.1. Mechanical dryer

Drying experiments were carried out at Bucaramanga, Colombia, the capital of Santander Department, which is the major cocoa producer of this country. The experimental facilities are located at the Universidad Pontificia Bolivariana Bucaramanga, specifically at the Automation and Control for Agro-industrial Processes Laboratory. Two automated units were used in this study; the first one is a conditioning device allowing to control humidity and temperature of highs amounts of air, developed to evaluate the performance of experimental dryers [33]. The second one is a rotary dryer equipped with an electric heater and a set of infrared lamps. It features temperature, humidity, speed and weight sensors, and automatic control of product's surface temperature and drum´s rotation speed. Figure 1 shows the aforementioned equipment with some of their components.

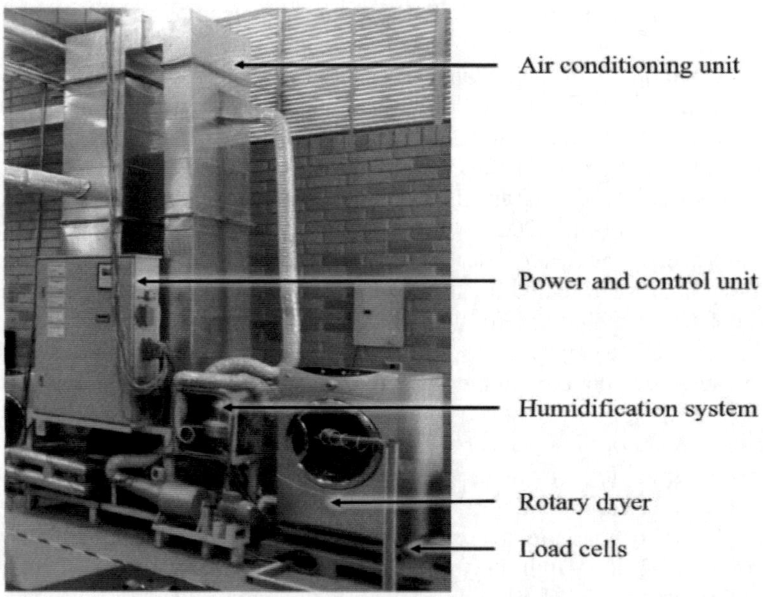

Figure 1. Air conditioning unit with rotary dryer

2.2. *Drying procedure*

Fresh cocoa pots of the Forastero variety were supplied by farmers of a production zone nearby (Lebrija, Santander). After harvesting, cocoa pots were transported to the drying facilities. Using a sharp knife cocoa shells were opened in two halves and then the cocoa beans were manually extracted. Afterwards, fermentation of 10 kg of the later was then done in a cylindrical wooden drum for 5 days while temperature of the product load was monitored. Cocoa beans were periodically turned over by rotating the drum. After fermentation, cocoa beans were loaded into the rotary dryer where three different drying configurations were set for the study. For all experiments temperature of the hot air was set at 50 °C, while surface temperature of cocoa beans was controlled by turning on and off the infrared lamps. Set points were stablished at 50 °C, 55 °C and 60 °C respectively. Drying was stopped when there were not significant changes in drum weight. Average time for the former treatments were 7.1 hours, 8.5 hours and 9.25 hours respectively.

2.3. Chemicals and reagents

All solvents were analytical or HPLC grade supplied by Panreac and Merck. Standard compounds, i.e. caffeine, theobromine, (+)-catechin, (-)-epicatechin were supplied by Fitolab, while the Gallic acid by Merck. The Folin & Ciocalteu's reagent was obtained from Merck. Deionized water was obtained with a Synergy UV ultrapure (Type 1) system.

2.4. Samples preparation

Dried beans were grinded using an electric milling machine just before a 200μm mesh size was attained. Some small pieces of solid carbon dioxide were mixed with the beans to avoid heat generation. Subsequently, 100 mg of this powder was degreased by adding 2 mL of hexane, centrifuged at 1565 g for 10 minutes, decanted and then, another 2 mL of hexane was added. This procedure was repeated three times. Afterwards, the fat reduced cocoa powder was extracted at room temperature with 1.5 mL aqueous 2-propanol (60%, pH 9.0) for 60 min in an ultrasonic bath. The mixture was centrifuged for 10 minutes at 391 g at 4°C and the supernatant was decanted and filtered through a nylon membrane (0.45 μm). The supernatant was poured into a flask to obtain 2mL of extract. This extract was stored at -20°C if not used immediately.

2.5. Determination of polyphenol content

Total polyphenols content (TCP) of extracts was determined spectrophotometrically using a method proposed by Londoño et al. [34]. First, it was prepared a mixture of 30 μm of the later extract, 30 μm of 20% Na_2CO_3, 225 μm of distilled water and15 μm of Folin & Ciocalteu's reagent. Then, one hour later the absorbance of the blue was measured at 760 nm against a blank sample and a calibration curve with gallic acid as standard was produced. The TCP is then expressed as milligram equivalent of gallic acid (mg GAE) per gram of dried sample material. All measurements were run in triplicate.

2.6. HPLC Analysis of Catechins and Methylxanthines

Analysis of cathechins and methylxanthines were performed with a Flexar HPLC System (Perkin Elmer Technology) featuring an UV detector, a vacuum degasser, an auto-sampler and a quaternary pump. Before HPLC analysis, samples were centrifuged for 15 minutes at 10581 g and 4°C. After this, it was filtered with a 0.45 μm nylon membrane.

To separate the compounds, the flow rate used varied between 0.75 and 1.0 mL/min and the analysis was carried out at 30 °C. A Zorbax Eclipse Plus C18 (25 cm × 4.6 mm, with a 5 μm particle size) column was used for the chromatographic separation. Water with 0.1% acetic acid as eluent A and methanol as eluent B were used as mobile phase. Due to variety of components, to obtain an ideal resolution of all peaks, it was followed an isocratic elution, using: 5% A and 95%B, for 10 minutes, with wavelength between 200 and 400 nm. Polyphenols ((+)-catechin and (-)-epicatechin) and Methylxanthines (theobromine and caffeine) were identified by comparing the retention times and spectral data to those of standards. All analyses were run in triplicate. Quantitative interpretation of chromatograms involved a comparison with an equal volume standard; this method is useful to determine the concentration of a single component and require analyze duplicates to guarantee system reliability [35].

2.7. *Statistical Analysis*

TPC and concentrations of the Methylxanthines and Polyphenols compounds were statistically evaluated with MS Excel, and a completed randomized design was applied with three replications. Data are presented then as means ± standard deviation.

3. Results and discussion

3.1. *Total polyphenols content (TPC).*

Figure 2 shows TPC of cocoa samples expressed in mg GAE/g dry cocoa for each treatment. Here, surface temperature of grains was imposed by controlling the FIR power input, while air temperature was fixed at 50°C by the heating and humidification unit.

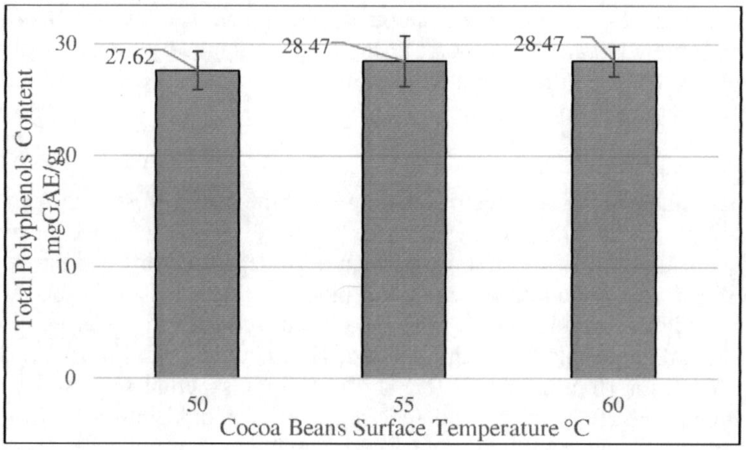

Figure 2. TPC as a function of cocoa beans surface temperature. Data are presented as mean±SD (n=3)

These averaged values indicate that polyphenols content basically does not change when beans surface temperature is increased. However, these results are lower than those obtained for row samples of Amazon and Forastero cultivar from Equator, Colombia and Ivory Coast. (84 mg GAE g^{-1}, 81 mg GAE g^{-1} and 81.50 respectively)[36]. There is only one study that obtained similar concentrations in samples of the Trinitarian cultivar -approximately 29 mgGAEg^{-1}-. In this case, cocoa beans were dried 20 hours with air at 50°C. Additionally, it was also found a concentration of about 13 mgGAEg^{-1} when air temperature was 60°C [37].

In this same framework, a study on commercially available sun-dried samples of Colombian cocoa beans (from 11 different geographic areas), obtained TPC values ranging between 44.940 ± 1.174 mgGAEg-1 and 70.090 ± 1.988 mgGAEg-1[24]. Likewise, sun dried cocoa beans samples from Venezuela yielded a TPC of 66.6±0.4 mgGAEg-1[38]. Besides, samples from Malaysia were also assessed under two different experimental approaches. On the one hand, cocoa beans were dried in a convection oven for 24 hours with air at 60°C, registering a TPC of 42.1±0.4 mg GAE g^{-1}[39]. On the other hand, cocoa beans were processed in a convection oven with relative humidity control, showing a TPC of 34.7 ±0.9 mgGAEg-1 after 40 hours of drying at 80°C.[40].

3.2. *Caffeine, theobromine, epicatechin and catechin content*

Quantification of the methylxanthines and polyphenols compounds of interest were performed by HPLC. Accordingly, standard curves for caffeine, theobromine, (-)-epicatechin and (+)-catechin were obtained with standards in concentrations between 10ppm and 60ppm, following the method described in section 2.6. The equation standards derived to quantify the molecules in the samples are displayed in table 1.

Table 1. Equation standards

Standards	Equation	R^2
(+)-catechin	Y= 1392.4X - 7850.5	0.9918
(-)-epicatechin	Y = 1374.5X + 6968.9	0.9942
Theobromine	Y = 26446X + 87767	0.9949
Caffeine	Y = 4250.2X + 55001	0.9971

Y= Area under the curve; X= compound(ppm).

In this study, methanol and diluted acetic acid as solvent allowed the identification of the former compounds by their retention times, HPLC-UV spectra (280 nm), and chromatographic comparisons with primary standards. Figure 3 shows a chromatogram for one of the samples of this study. Results in general showed no qualitative differences in chromatographic profiles between cocoa extracts of all treatments. However, there are quantitative differences in their concentrations as is shown in Table 2.

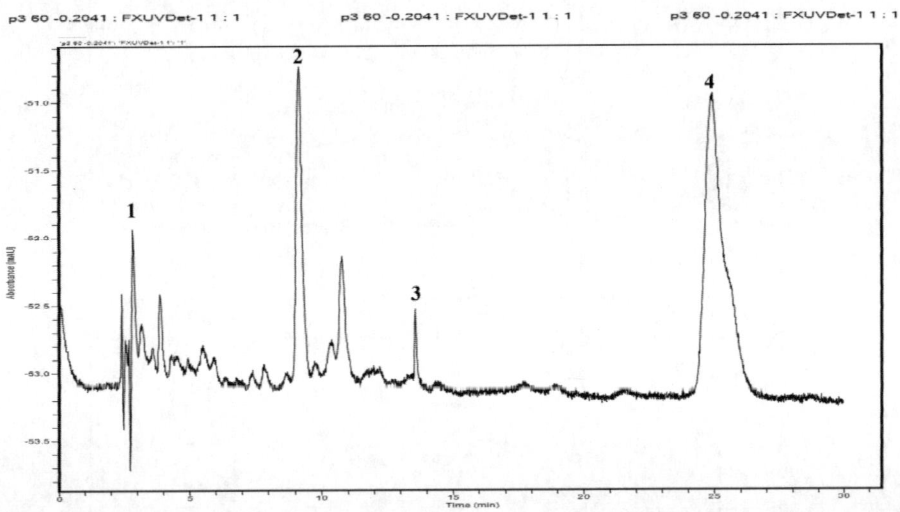

Figure 3. HPLC chromatographic profiles for cocoa bean extracts at 60 °C. Chromatographic peaks for theobromine (1), (+)- catechin (2), caffeine (3) and (-)-epicatechin (4); with retention times of 3.88 min, 9.12 min, 13.5 min and 25 min respectively.

Chromatograms also revealed the evolution of new peak concentrations as surface temperature of cocoa beans was increased. That is, while concentrations of (+)-catechin and (-)-epicatechin peaks decreases (see figure 4), there is a rise of peaks to the right of the former analytes that could suggest the formation of new molecules. Nonetheless, caffeine picks are stable at all surface temperatures.

Regarding chemical species concentrations, results agree well with other previous studies. Theobromine amounts are much higher than the other compounds as it has been already founded in most of the previous cocoa studies. However, it is particularly higher, especially for cocoa beans at 50°C surface temperature[24]. Furthermore, with the increase of beans surface temperature, concentration of all compounds is lower. Still, total polyphenols content (mg GAE/g) keeps almost constant with bean surface temperature, which could be explained by the presence of other phenolic compounds like procyanidins [36,41] or new compounds resulting from thermally induced epimerization of (-)-epicatechin and (+)-catechin [42,43].

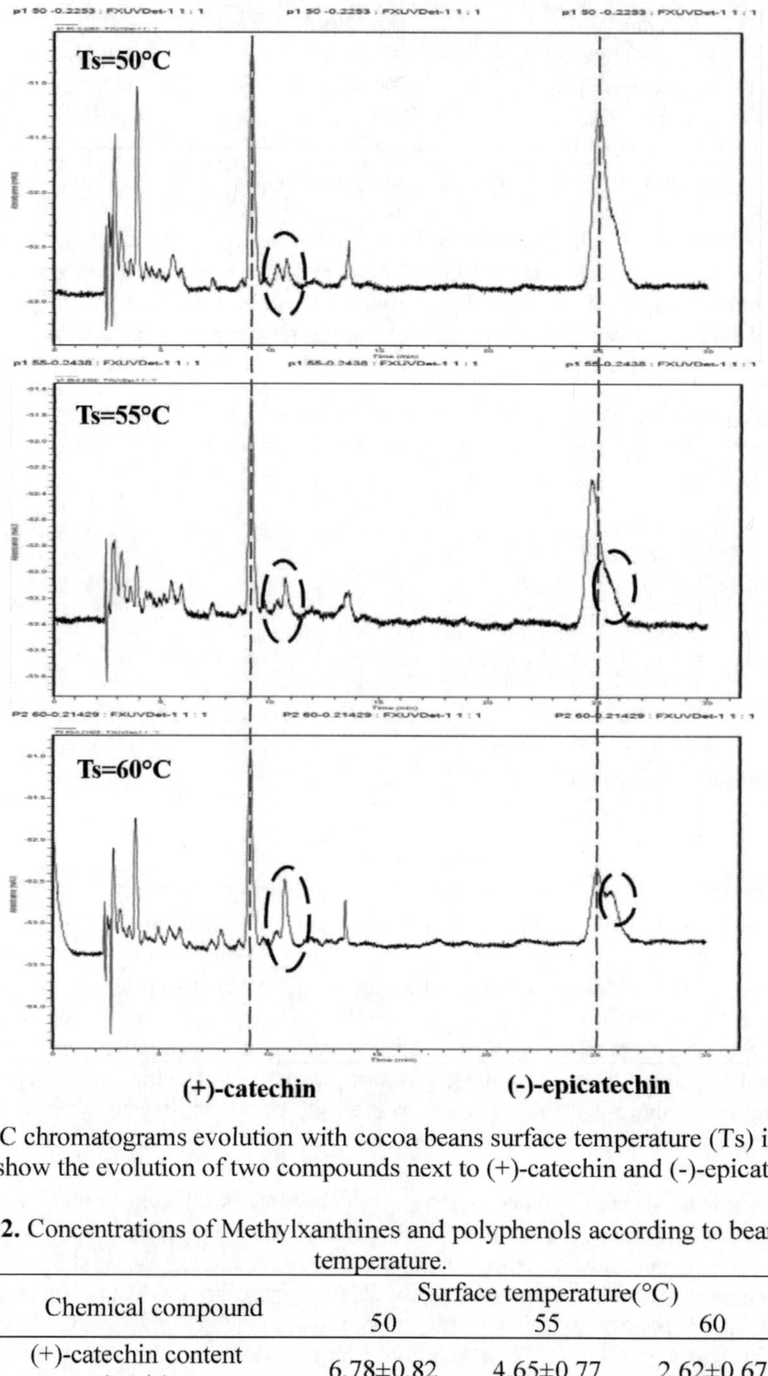

Figure 4. HPLC chromatograms evolution with cocoa beans surface temperature (Ts) increase. Doted areas show the evolution of two compounds next to (+)-catechin and (-)-epicatechin.

Table 2. Concentrations of Methylxanthines and polyphenols according to beans surface temperature.

Chemical compound	Surface temperature(°C) 50	55	60
(+)-catechin content (mg/g)	6.78±0.82	4.65±0.77	2.62±0.67
(-)-epicatechin content (mg/g)	3.84±0.008	3.84±0.013	2.02±0.023
Theobromine content (mg/g)	21.66±0.56	15.06±1.26	12.99±1.70
Caffeine (mg/g)	1.38±0.12	0.80±0.06	0.61±0.023

Data are presented as mean ± SD (n=3).

3.2.1. Theobromine and caffeine ratio

This ratio has been used to identify cocoa cultivars. According to former studies, ratios over 9 normally correspond to Forastero samples and ratios under 3 are correlated to Criollo ones. Trinitario variety normally yields ratios between 3 and 9 [24]. The calculated values in this study (Figure 5) are similar to those reported for diverse cacao samples from Peru [44].

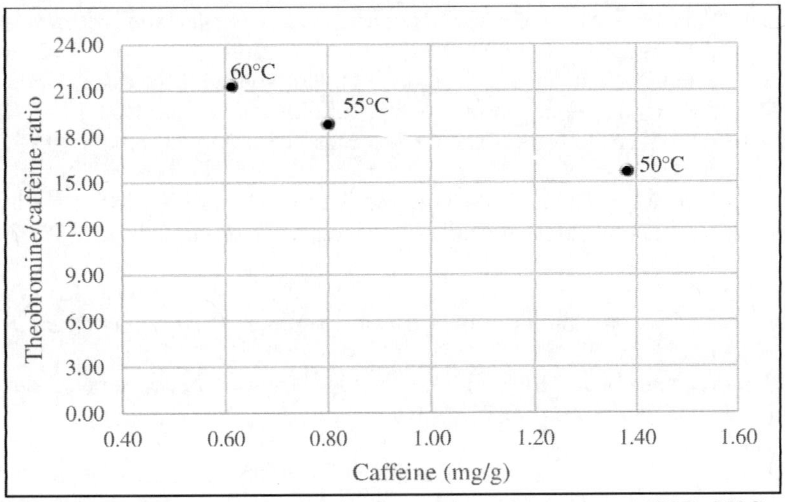

Figure 5. Ratio Theobromine/caffeine for the evaluated samples at 50°C, 55°C and 60°C.

4. Conclusions

Bearing in mind that combined convective and infrared drying for cocoa beans has not been well studied yet, this research work has revealed its potential to support cocoa postharvest processes. On one hand, the analysis of cocoa samples dried under this combined strategy in the rotary unit, indicates that chemical compounds like Methylxanthines and Polyphenols can be founded in similar amounts than those in fresh cocoa and sun-dried cocoa beans. This fact added the use of a mechanical dryer can improve overall quality of cocoa beans, allowing control of the process and reducing drying times. Results also indicate that TPC does not essentially change when the cocoa beans surface temperature in the dryer was increased from 50°C to 60°C.

This experimental work also suggest that it is required a more comprehensive study to weigh the effect of other variables like cocoa cultivars, drying air temperature, cocoa beans surface temperature and drying time not only on the chemical composition of the samples but also on other quality criteria related to organoleptic evaluation and colour assessment. It is also worth using HPLC or similar analytical techniques to assess the presence of other active polyphenol compounds in the dried samples and evaluate the impact of infrared radiation on their respective concentrations.

Acknowledgments

The authors gratefully acknowledge the financial support from the Research and Transfer Department (DIT) and the Faculties of Mechanical Engineering and Environmental Engineering of the Universidad Pontificia Bolivariana Seccional Bucaramanga.

References

[1] García-Cáceres R G, Perdomo A, Perdomo A, Ortiz O, Ortiz O, Beltrán P, Beltrán P, López K and López K 2014 Characterization of the supply and value chains of Colombian cocoa *DYNA* **81** 30–40

[2] Rojas F and sacristán E 2013 Guia ambiental para el cultivo de cacao

[3] Jinap S, Dimick P S and Hollender R 1995 Flavour evaluation of chocolate formulated from cocoa beans from different countries *Food Control* **6** 105–10

[4] Afoakwa E O, Paterson A, Fowler M and Ryan A 2008 Flavor formation and character in cocoa and chocolate: A critical review *Critical Reviews in Food Science and Nutrition* **48** 840–57

[5] Arlorio M, Locatelli M, Travaglia F, Coïsson J-D, Grosso E D, Minassi A, Appendino G and Martelli A 2008 Roasting impact on the contents of clovamide (N-caffeoyl-L-DOPA) and the antioxidant activity of cocoa beans (Theobroma cacao L.) *Food Chemistry* **106** 967–75

[6] Caligiani A, Cirlini M, Palla G, Ravaglia R and Arlorio M 2007 GC-MS detection of chiral markers in cocoa beans of different quality and geographic origin *Chirality* **19** 329–334

[7] Counet C, Ouwerx C, Rosoux D and Collin S 2004 Relationship between Procyanidin and Flavor Contents of Cocoa Liquors from Different Origins *J. Agric. Food Chem.* **52** 6243–9

[8] Hansen C E, del Olmo M and Burri C 1998 Enzyme activities in cocoa beans during fermentation *J. Sci. Food Agric.* **77** 273–81

[9] Hashim P, Selamat J, Muhammad K and Ali A 1999 Effect of drying time, bean depth and temperature on free amino acid, peptide-N, sugar and pyrazine concentrations of Malaysian cocoa beans *J. Sci. Food Agric.* **79** 987–94

[10] Wollgast J and Anklam E 2000 Review on polyphenols in Theobroma cacao: changes in composition during the manufacture of chocolate and methodology for identification and quantification *Food Research International* **33** 423–47

[11] Bharath S and Bowen-O'Connor C 2008 Assessing drying rates of cacao beans using small samples *Cocoa Research Unit The University of the West Indies* 52

[12] Faborode M O, Favier J F and Ajayi O A 1995 On the effects of forced air drying on cocoa quality *Journal of Food Engineering* **25** 455–72

[13] Hii C L, Abdul Rahman R, Jinap S and Che Man Y 2006 Quality of cocoa beans dried using a direct solar dryer at different loadings *J. Sci. Food Agric.* **86** 1237–43

[14] Jinap S, Thien J and Yap T N 1994 Effect of drying on acidity and volatile fatty acids content of cocoa beans *J. Sci. Food Agric.* **65** 67–75

[15] Rodriguez-Campos J, Escalona-Buendía H B, Contreras-Ramos S M, Orozco-Avila I, Jaramillo-Flores E and Lugo-Cervantes E 2012 Effect of fermentation time and drying temperature on volatile compounds in cocoa *Food Chemistry* **132** 277–88

[16] Zahouli G I B, Guehi S T, Fae A M, Ban-Koffi L and Nemlin J G 2010 Effect of drying methods on the chemical quality traits of cocoa raw material *Advance Journal of Food Science and Technology* **2** 184–90

[17] Bonaparte A, Alikhani Z, Madramootoo C A and Raghavan V 1998 Some quality characteristics of solar-dried cocoa beans in St Lucia *J. Sci. Food Agric.* **76** 553–8

IOP Conf. Series: Materials Science and Engineering **437** (2018) 012011 doi:10.1088/1757-899X/437/1/012011

[18] García-Alamilla P, Salgado-Cervantes M A, Barel M, Berthomieu G, Rodríguez-Jimenes G C and García-Alvarado M A 2007 Moisture, acidity and temperature evolution during cacao drying *Journal of Food Engineering* **79** 1159–65

[19] Ndukwu M C 2009 Effect of Drying Temperature and Drying Air Velocity on the Drying Rate and Drying Constant of Cocoa Bean *Agricultural Engineering International: CIGR Journal* **0**

[20] Hii C L, Law C L and Cloke M 2009 Modeling using a new thin layer drying model and product quality of cocoa *Journal of Food Engineering* **90** 191–8

[21] Manoj M and Manivannan A 2013 Simulation of solar dryer utilizing greenhouse effect for cocoa bean drying *IJAET* **IV** 24–7

[22] Saltini R, Akkerman R and Frosch S 2013 Optimizing chocolate production through traceability: A review of the influence of farming practices on cocoa bean quality *Food Control* **29** 167–87

[23] Humston E M, Knowles J D, McShea A and Synovec R E 2010 Quantitative assessment of moisture damage for cacao bean quality using two-dimensional gas chromatography combined with time-of-flight mass spectrometry and chemometrics *Journal of Chromatography A* **1217** 1963–70

[24] Carrillo L C, Londoño-Londoño J and Gil A 2014 Comparison of polyphenol, methylxanthines and antioxidant activity in Theobroma cacao beans from different cocoa-growing areas in Colombia *Food Research International* **60** 273–80

[25] Cooper K A, Campos-Giménez E, Alvarez D J, Nagy K, Donovan J L and Williamson G 2007 Rapid reversed phase ultra-performance liquid chromatography analysis of the major cocoa polyphenols and inter-relationships of their concentrations in chocolate *Journal of Agricultural and Food Chemistry* **55** 2841–7

[26] Kyi T M, Daud W R W, Mohamad A B, Samsudin M W, Kadhum A A H and Talib M Z M 2005 The kinetics of polyphenol degradation during the drying of Malaysian cocoa beans *International Journal of Food Science and Technology* **40** 323–31

[27] Rodriguez-Campos J, Escalona-BuendÃ-a H B, Orozco-Avila I, Lugo-Cervantes E and Jaramillo-Flores M E 2011 Dynamics of volatile and non-volatile compounds in cocoa (Theobroma cacao L.) during fermentation and drying processes using principal components analysis *Food Research International* **44** 250–8

[28] Pura Naik J 2001 Improved high-performance liquid chromatography method to determine theobromine and caffeine in cocoa and cocoa products *Journal of Agricultural and Food Chemistry* **49** 3579–83

[29] Frauendorfer F and Schieberle P 2008 Changes in key aroma compounds of Criollo cocoa beans during roasting *Journal of Agricultural and Food Chemistry* **56** 10244–51

[30] Caligiani A, Acquotti D, Cirlini M and Palla G 2010 1H NMR study of fermented cocoa (Theobroma cacao L.) beans *Journal of Agricultural and Food Chemistry* **58** 12105–11

[31] Caligiani A, Palla L, Acquotti D, Marseglia A and Palla G 2014 Application of 1H NMR for the characterisation of cocoa beans of different geographical origins and fermentation levels *Food Chemistry* **157** 94–9

[32] Hii C L, Law C L, Suzannah S, Misnawi and Cloke M 2009 Polyphenols in cocoa (Theobroma cacao L.). *Asian Journal of Food and Agro-Industry* **2** 702–22

[33] Hernández B D O, Olejua A R C and Olarte J M C 2016 Automatic humidification system to support the assessment of food drying processes *IOP Conf. Ser.: Mater. Sci. Eng.* **138** 12019

[34] Londoño L J, Montoya P G, Guerrero M K, Aristizabal L, A A and J G 2006 Citrus juice inhibits low density lipoprotein oxidation: relationship between free radical scavenger activity and electrophoretic mobility *Revista chilena de nutrición* **33** 544–51

[35] Yost R W ; L S E R D C 1980 *Practical Liquid Chromatography: an Introduction* (Perkin-Elmer)

[36] Oracz J, Zyzelewicz D and Nebesny E 2015 The Content of Polyphenolic Compounds in Cocoa Beans (*Theobroma cacao* L.), Depending on Variety, Growing Region, and Processing Operations: A Review *Critical Reviews in Food Science and Nutrition* **55** 1176–92

[37] Alean J, Chejne F and Rojano B 2016 Degradation of polyphenols during the cocoa drying process *Journal of Food Engineering* **189** 99–105

[38] Padilla F C, Rincón A M and Bou-Rached L 2008 Contenido de polifenoles y actividad antioxidante de varias semillas y nueces *Archivos Latinoamericanos de Nutrición* **58** 303–8

[39] Teh Q T M, Tan G L Y, Loo S M, Azhar F Z, Menon A S and Hii C L 2016 The Drying Kinetics and Polyphenol Degradation of Cocoa Beans *Journal of Food Process Engineering* **39** 484–91

[40] Abhay S M, Hii C L, Law C L, Suzannah S and Djaeni M 2016 Effect of hot-air drying temperature on the polyphenol content and the sensory properties of cocoa beans. *International Food Research Journal* **23** 1479–84

[41] Patras M A, Milev B P, Vrancken G and Kuhnert N 2014 Identification of novel cocoa flavonoids from raw fermented cocoa beans by HPLC–MSn *Food Research International* **63** 353–9

[42] Jalil A, Ismail A, Jalil A M M and Ismail A 2008 Polyphenols in Cocoa and Cocoa Products: Is There a Link between Antioxidant Properties and Health? *Molecules* **13** 2190–219

[43] Park J-H, Lee J-M, Cho Y-J, Kim C-T, Kim C-J, Nam K-C and Lee S-C 2009 Effect of Far-Infrared Heater on the Physicochemical Characteristics of Green Tea During Processing *Journal of Food Biochemistry* **33** 149–62

[44] Peláez P P, Bardón I and Camasca P 2016 Methylxanthine and catechin content of fresh and fermented cocoa beans, dried cocoa beans, and cocoa liquor *Scientia Agropecuaria* **7** 355–65

IOP Conf. Series: Materials Science and Engineering **437** (2018) 012012 doi:10.1088/1757-899X/437/1/012012

Possibilities of carbon black recovery from waste tyre pyrolysis to be used as additive in rubber goods -a review-

N Cardona[1], F Campuzano[1], M Betancur[1], L Jaramillo[2] and J D Martínez[1]

[1]Grupo de Investigaciones Ambientales, Universidad Pontificia Bolivariana, Circular 1ra No. 70-01 Bloque 11, Medellín, Colombia
[2]Grupo de Investigación en Calidad, Metrología y Producción, Instituto Tecnológico Metropolitano, Cra. 74d #732, Medellín, Colombia

natalia.cardonaur@upb.edu.co

Abstract. This work deals with the literature review of the main carbon black (CB) properties required for reinforcing-filler rubber applications. Commercial CB is rich-carbon material widely used as reinforcing filler in rubber goods. When waste tyres are transformed by pyrolysis, the obtained solid fraction known as pyrolytic carbon black (CBp) contains the original CB (80-90 wt.%) added in tyre manufacturing process along with sulphur (1-3 wt.%) and a representative quantity of inorganics (10-15 wt.%). Taking this into account, it could be inferred that some physical-chemical properties of CBp are similar to those of the original CB. This fact has encouraged the use of CBp as an alternative substitute of CB in different rubber applications. However, it has been found that CBp can be used in certain rubber applications, some research works have also noted the detrimental of some mechanical properties. The inorganics presence in CBp limits some important properties such as tensile strength, tear strength, and hardness in the rubber composites. Also, these compounds as well as some carbonaceous deposits seem to affect the surface area and the surface chemistry, and therefore the active sites, affecting the reinforcing activity. To overcome these limitations, some post-treatments processes are proposed.

1. Introduction

The accelerate growth of the populations as well as the remarkable increase of vehicles worldwide have led to an environmental and public concern over the waste tyres (WTs) disposal in the last few years. It is forecasted that the growth of tyres demand all over the world is about 4.3% per year reaching 2.9 billion units in 2017 [1]. Tyres show several problems at the time to be disposed since they are designed to be extremely resistant to physical, chemical and even biological degradation. Moreover, their storage represents a big risk of fire that would cause severe environmental impacts due to uncontrolled toxic gaseous emissions as occurred in Madrid (Spain) in 2015 [2]. To overcome these issues, different recycling alternatives have been proposed over the years, including retreating, and material and energy recovery practices.

The progress made in recent years in the management of polymer wastes has made WTs to be perceived as valuable materials [3]. Considering that tyres are composed by natural and synthetic rubber (60-65 wt.%), carbon black (25-35 wt.%) and inorganic additives, it is reasonable to find routes for energy and materials recovery purposes[4]. In this sense, pyrolysis of WTs has been considered one of the most promising pathways over other possible valorisation methods such as combustion and gasification. This is due to its less environmental impact and the possibility to recover liquid and solid compounds with a wide potential in the energy and material sectors [5–7].

In the WTs pyrolysis process, the solid fraction is, along the liquid fraction, the most abundant obtained fraction since it considers around 30-40 wt.% of the tyre weight [8]. This fraction, also known as pyrolytic carbon black (CBp), is composed of a mixture of commercial carbon blacks (CB) which were added in the tyre manufacturing process, carbonaceous deposits (as a result of secondary reactions) and some inorganics such as Zn, Ca, and Si added during tyre manufacture. This composition makes CBp to be different from the commercial CB which has limited its application and profitability in the current market. However, it has been argued that the careful control of the feedstock and pyrolysis processing conditions may lead to CBp grades specially tailored for certain applications [9]. For instance, one of the most immediate uses of CBp is the production of activated carbon as reviewed by Mui et al [10]. Similarly, low-grade rubber pieces and raw material for both pavements and pigments, among others, have been reported to be a technical and a feasible way to use this material. On the other hand, it is also possible use the CBp as a solid fuel despite its slow oxidation kinetics [11]. Besides, some demineralization processes have been also conducted to remove the inorganic material as well as the carbonaceous deposits in order to reach CBp with similar properties to those ones of commercial CBs [12, 13].

Conventionally, CB is usually produced by the partial oxidation of hydrocarbons leading to high CO_2 and polycyclic aromatic hydrocarbons (*PAHs*) emissions that not only affect the environment but also the human health [14, 15]. For instance, the incomplete combustion process carried out to obtain the commercial CB drives to a carbon footprint around 5.7 kg of equivalent CO_2 per kg of CB [16]. Indeed, it has been recognized that the carbonaceous particulate matter emitted to the environment during this process has a global warming potential between 330 and 2240 kg of equivalent CO_2 [17]. Likewise, the International Agency of Research on Cancer (IARC) has defined *PAHs* as carcinogenic substances [18]. These facts have motivated the idea of recovering CB from WTs through pyrolysis processes. In this sense, the obtained CBp could be reused in new rubber goods as well as be precursor of other carbonaceous materials [19]. In spite of the difference between CB and CBp mainly associated with the high concentration of inorganic impurities and carbonaceous deposits, the CBp could present a big potential for its application in rubber formulations as a partial substitute as will be shown later.

Taking this into account, the present work deals with the literature review of the CBp properties and some scientific results when it is used in different rubber goods. According to the best authors' knowledge, there is not this kind of study in the scientific literature, which represents an important contribution not only to the problems associated with WTs disposal but also to substitute the commercial CB given the environmental concerns related to its production.

2. Commercial carbon black (CB)

Commercial CB is essentially elemental carbon (higher than 97 wt.%) in the form of extremely fine particles with amorphous molecular structure. It is produced by the incomplete combustion or thermal decomposition of gaseous or liquid hydrocarbons under controlled conditions. Even though CB is mainly used as reinforcing filler in the rubber industry (90%), it has also other important applications such as pigments for printing inks (9%) and as an additive in hundreds of diverse processes (1%) [20]. When this material is used in rubber goods (especially tyre manufacture), it is added to improve the processing, strength and durability of elastomer composites [21]. Some CB production processes include oil furnace, thermal, lamp, acetylene cracking, gas furnace and channel processes. Nevertheless, furnace and thermal accounts for about 90% and 10% of CB production in U.S., respectively. Besides, the lamp and acetylene processes are small-volume specialty black operations that constitute less than 1% of total production in U.S. Whereas, the gas process is being phased out and the last channel black plant in the U. S. was closed in 1976 [22].

2.1. Oil Furnace black process

The oil furnace black or simply furnace black process is based on the incomplete combustion of heavy hydrocarbons. This process is carried out in a closed reactor with atomizing nozzles [20]. The fuel is sprayed into a hot gas stream generated by burning a secondary feedstock and pre-heated [3, 20]. Then, the hydrocarbons drops are vaporized and then pyrolyzed to give rise to carbon particles. Finally, the cracking reaction is stopped by adding steam or liquid water sprays. The formed CB flows from the reactor through heat exchanger where it is cooled down and then collected in bag filters [20]. Process yields range from 35 to 65 percent, depending on the feed composition and the grade of CB desired. Generally, yields are highest for large particle blacks and lowest for small particle CB [22]. This process has become the predominant method for CB production since the hydrocarbons serve both as a source of heat and carbon [3].

2.2. Thermal black process

This process consist in a thermal decomposition of natural gas in the absence of air [18, 23]. The system consists in two refractory-line furnaces that alternate between furnace preheating and CB production approximately every five minutes. The natural gas is injected into one previously preheated refractory-line furnace (about 1300 °C [24]) in air absence, while the other furnace remains in the heating cycle. Then, the heat from the refractory material in the furnace decomposes the natural gas into CB and H_2. Finally, the aerosol material is cooled down with water sprays and collected in bags. The H_2 is burned using air to pre-heat the reactor in the heating cycle for later production cycle [20, 22]. In fact, more than enough H_2 is produced to make the thermal black process self-sustaining, and the excess H_2 is used to produce steam and electric power [22].

Generally speaking, commercial CB is classified by ASTM into different classes such as N120, N330, N550 and N660. Herein, the N is used to indicate a normal curing rate typical of furnace blacks that have received no special modification to alter their influence on the rate of cure of rubber. The next digits indicate the typical average particle size [18, 20]. Taking this into account, the higher the CB reference number, the lower the specific surface area (e.g. for reinforcing applications big surface areas are required). Moreover, thermal black is classified into the N900 series while furnace blacks fall within the N100 to N700 series [25]. The most common CBs used in tyre manufacturing are N330 and N550, which present a specific surface area of 78 and 40 m^2g^{-1}, respectively. These references are used in low hysteresis tyre treads, inner liner tyres, carcass, sidewall, among others. In addition, due to their high dimensional stability, these CBs are also used in hoses, belts plastic piping, etc [18, 26]. Table 1 shows the classification as well as some properties and applications of commercial CBs.

IOP Conf. Series: Materials Science and Engineering **437** (2018) 012012 doi:10.1088/1757-899X/437/1/012012

Table 1. Properties and applications of some commercial carbon blacks, adapted from [18, 26]

ASTM Classification	Iodine absorption (g kg^{-1})	Area BET (m^2g^{-1})	STSA (m^2g^{-1})	DBPA (mL100g $^{-1}$)	Ash (wt. %)	Sieve residue 45 µm (wt. %)	Density (kgm^{-3})	pH	Type	Production process	Average primary particle size (nm)	Reinforcing grade and applications
N110	145	127	115	113	n.r	n.r	n.r	n.r	SAF	Furnace	17	High reinforcement; high abrasion resistance; used in special and off-road tyre products.
N220	121	114	106	114	<0.5	<0.05	355	9	ISA F	Furnace	21	High reinforcement and tear strength, good processing; used in passenger, off-road and special tyres.
N330	82	78	75	102	<0.5	<0.05	380	7	HAF	Furnace	31	Medium-high reinforcement. Low modulus, high elongation, good flex, tear and fatigue resistance; used in tyre tread, carcass and sidewall compound, bicycle tyres.
N550	43	40	39	121	<1.5	<0.05	360	n.r	FF	Furnace	53	Medium-high reinforcement, high modulus and hardness, low die swell and smooth extrusion; used in tyre inner liners, carcass and sidewall compounds and hose and other extruded goods.
N660	36	35	34	90	<0.5	<0.05	440	n.r	GPF	Furnace	63	Medium reinforcement and modulus, good flex and fatigue resistance, low heat build-up; used in tyre inner liners, carcass and sidewall, sealing rings, cable jackets, hose and extruded goods.
N762	27	29	28	65	n.r	n.r	n.r	n.r	SRF	Furnace	110	Medium reinforcement, high elongation and resilience, low compression set; used in mechanical goods, footwear, inner tubes and floor mats.
N774	29	30	29	72	<0.5	<0.05	n.r	n.r		Furnace	n.r	Medium reinforcement, high loading capacity and low hysteresis. Easy extrusion black. Used in belts, hoses, molded goods, and footwear.
N990	n.r	8	8	43	<0.7	<0.1-0.05	355	7	MT	Thermal	320	Low reinforcement, low modulus, hardness, hysteresis and tensile strength, high elongation and loading capacity; used in wire insulation and jackets, mechanical goods, footwear, belts, hose, gaskets, O-rings and tyre inner liners.

SAF superabrasion furnace black, ISAF intermediate superabrasion furnace black, HAF high abrasion furnace black, FF fast extruding furnace black, GPF general purpose furnace black, SRF semi-reinforcing furnace black, MT medium thermal black, STAS statistical Surface area, DBPA: dibutyl phthalate absorption

3. Pyrolytic carbon black (CBp)

As mentioned before, CBp is a heterogeneous carbonaceous material obtained from the WTs pyrolysis process. Physically, CBp is visible as solid, black particles of varying size, depending on the WT size used during pyrolysis. CBp particles seem to be much coarser in size compared to their original CB counterparts present in tyres since CBp particles act as nuclei during pyrolysis favoring the growth of solid carbon structures by crosslinking or cyclisation [27]. This material typically contains a mixture of different commercial CBs (since each part of the tyre requires a specific CB reference), some carbonaceous deposits [28], and inorganics (SiO_2, ZnO, K_2O, Fe_2O_3, etc) coming from the tyre manufacturing process which are reflected in the ash content. The carbonaceous deposits are the consequence of repolimerization reactions (secondary reactions) that take place among the polymer-derivatives. In other words, some of the organic vapours released during pyrolysis are converted into coke by dealkylation and dehydrogeneation reactions or absorbed onto the CBp surface reducing its activity and morphology [29]. For instance, Darmstadt et al. [28] reported that the average size of the aromatic compounds onto the CBp surface is around five condensed rings when it is obtained at atmospheric conditions. However, some studies have demonstrated that low pressure (~20 kPa) and high temperature (up to 700 °C) during pyrolysis can reduce these carbonaceous deposits resembling some commercial CB (N115 to N774) [30, 31]. Likewise, the inorganic compounds seem to be absorbed onto the active sides of the CB reducing surface activity.

On the other hand, CBp shows a heating value between 25 and 34 MJ/kg and may have composition of carbon and sulphur higher than 80 and 2 wt.%, respectively. During pyrolysis, approximately 70 wt.% of the sulphur contained in the tyre (added during vulcanization) remains into the solid fraction in the form of sulphides [32]. Moreover, CBp exhibits a specific surface area between 57 and 89 m^2g^{-1}, which is comparable with the surface area of the commercial CBs N330 and N550, as showed in Table 1. In addition, Table 2 summarizes the elemental and proximate analyses as well as the higher heating value (HHV) and surface area (BET) of some CBp reported in the literature. The difference in ash content arises from the difference of tyre type (e.g. passenger, truck, bicycle, etc.) although it usually remains between 6 and 16 wt.% [33, 34]. Table 3 shows some CBp ashes composition reported in the literature. As observed, the main compounds are SiO_2 and ZnO which are used as filler and activating agent in tyre manufacturing process, respectively.

Roughly speaking, when CBp is considered as alternative to commercial CB, the main properties to take into account are particle size, structure, surface area and surface chemistry. The reinforcement mechanism between the CBp and the polymer is mainly influenced by the size of CBp aggregates (formed by primary particles strong linked), and the surface morphology and chemistry. Chemical and physical interactions lead to adsorption of polymer chains onto the CBp surface. These interactions restrict the segmental movement of the polymer molecules, giving place to different reinforcement grades. Increasing surface area and activity will increase polymer-filler interactions [9]. Taking this into account, it is a fundamental step to stablish these CBp properties and compare them with those of the commercial CB in order to define further applications.

Table 2. Proximate and elemental analyses of CBp reported in literature

η (wt.%)	Elemental analysis wt.% (a.r) C	H	N	S	Proximate analysis wt.% (a.r) M	A	VM	FC	HHV (MJ kg^{-1})	BET area (m^{2}g^{-1})	Temp. pyrolysis (°C)	Reactor type	Ref.
40.5	84.2	1.1	0.3	2.1	3.6	12.4	4.7	79.3	30.0	65.7	550	AR	[29]
37.9	82.7	0.4	< 1	2.2	n.r	n.r	n.r	n.r	n.r	69.5	500	FBR	[35]
37.6	n.r	n.r	n.r	n.r	0.0	16.1	5.5	78.4	27.9	n.r	550	FBR	[36]
39.9	80.8	1.5	0.5	2.4	1.3	14.6	6.9	77.2	30.0	89.1	550	RKR	[37]
36.1	85.2	1.2	0.6	2.0	n.r	13.3	n.r	n.r	n.r	68.0	500	FBR	[38]
38.0	90.3	0.3	0.2	1.2	0.1	8.4	0.7	90.8	n.r	63.0	500	FBR	[6]
40.0	86.3	0.3	0.3	2.8	0.4	12.5	1.8	91.3	29.7	64.0	550	FBR	[39]
33.0	80.1	0.4	0.2	2.8	1.0	16.5	1.2	81.3	28.6	n.r	550	FBR	[34]
41.7	n.r	n.r	n.r	1.0	n.r	40.8	n.r	55.9	15.7	60.5	650	FBR	[40]
33.8	n.r	n.r	n.r	2.3	n.r	14.8	n.r	82.6	33.4	63.5			
49.1	85.3	0.3	0.3	2.1	3.6	15.3	12.8	71.9	30.7	n.r	550	RKR	[41]
35.0	89.4	1.8	1.8	0.9	n.r	7.7	n.r	n.r	30.9	n.r	550	FBR	[42]
52.5	0.0	0.0	0.0	2.6	n.r	n.r	n.r	n.r	n.r	n.r	550	FBR	[43]
34.1	86.6	0.8	0.8	2.2	n.r	n.r	n.r	n.r	n.r	65.2	500	CSBR	[44]
36.9	87.4	0.4	0.4	3.3	n.r	n.r	n.r	n.r	n.r	77.9			
38.0	81.1	0.4	0.4	3.4	1.2	13.2	3.5	82.1	n.r	n.r	600	AR	[45]
n.r	81.8	0.3	0.3	3.0	0.3	13.8	2.5	83.4	n.r	n.r		FBR	
n.r	n.r	n.r	n.r	n.r	1.1	18.7	4.4	76.9	n.r	57.0	500	PDU	[46]
35.0	80.7	0.4	0.7	3.6	n.r	14.4	n.r	n.r	n.r	n.r	500 (20 kPa)	PDU	
35.0	87.2	0.4	0.5	2.0	n.r	9.9	n.r	n.r	n.r	n.r	500	PDU	[28]
35.0	85.7	1.0	0.4	1.3	n.r	9.1	n.r	n.r	n.r	n.r	500	Rotatory reactor	
n.r	87.4	0.7	0.3	2.6	0.2	9.5	2.6	77.7	n.r	82.0		FBR	[47]

n.r: not reported; M: moisture; A: ash; VM: volatile matter; FC: fixed carbon; C: elemental carbon; H: hydrogen; N: nitrogen; S: sulphur; HHV: Higher Heating value; η: CBp yield; FBR: Fixed Bed Reactor; AR; Auger Reactor; PDU: Process Development Unit, CSBR: Conical Spouted Bed Reactor, MBR: Moving Bed Reactor. RKR: Rotary Kiln Reactor.

Table 3. Ash composition of pyrolytic carbon black

Al_2O_3	CaO	Fe_2O_3	K_2O	MgO	Na_2O	SiO_2	TiO_2	ZnO	Total	Ref.
1.9	4.9	1.2	1	1.3	0.7	39	0.1	31.7	81.8	[29]
11.5	10.2	2.7	0.1	1	2.4	27	n.r	29.6	84.5	[48]
3.1	26	1.4	1.3	0.8	1	2.6	n.r	48.1	84.2	[49]
0.2	0.4	0.1	0.2	0.1	n.r	2.7	n.r	2.5	6.2	[33]

n.r: not reported

4. Main properties of both CB and CBp for rubber formulations

As aforementioned, the characteristics of CB and in this case of CBp are determined by some fundamental physical-chemical properties. The size, shape and distribution of aggregates are generally termed CBp structure. Likewise, CBp surface activity is described in terms of the oxygen containing chemical groups onto the surface (e.g carboxyl, quinone, phenol and lactone) and delocalized π electron [50]. The smaller the particle sizes (<50 nm), the higher the surface area and hence, the higher the reinforcement grade [18, 51, 52]. Generally, the higher the surface area, the surface activity and the structure, the higher the CB reinforcing potential.

4.1. Particle, aggregate and agglomerate

CB exhibits hierarchy morphological features: (i) particles (that is, primary particles), (ii) aggregates, and (iii) agglomerates, as shown in Figure 1. Even though the fundamental building block is based on primary particles, they tend to strongly linked by covalent bonds forming aggregates. Thereafter, individual aggregates get together by van der Waals forces to form agglomerates. These agglomerates do not break down into smaller components unless adequate forces are applied [20]. Particle size is an important property since it contributes to strength and blackness in rubber goods [53]. The average primary particle size of commercial rubber type CB grades ranges from about 10 to 500 nm [18, 54]. The thermal CB has the largest particle size (mean diameters of 240 – 320 nm) [18]. In contrast, the furnace CB shows particle sizes approximately three to twenty times smaller (mean diameters of 15 – 80 nm) [18, 25]. In the case of CBp, there is an irregular distribution of particle size given the several CB grades used in tyre manufacture. Huang et al. [31] reported that commercial CB aggregates have a Gaussian distribution. They found three different aggregates distribution for the CBp produced at 500°C in a tubular reactor: (i) 80-115 nm, (ii) 212-253 nm, and (iii) 418-474 nm, which is far from a single commercial CB reference. Likewise, Du et al. [55] noticed by scanning electron microscopy (SEM) that the aggregation size of CBp (~10 μm) is larger than that of commercial CB N774 (~2 μm) and similar to Huang et al., they observed wider aggregate distributions for CBp.

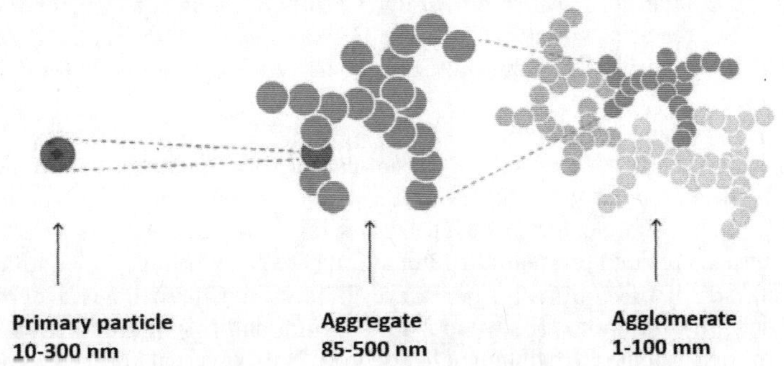

Figure 1. CB hierarchy morphological features [20]

4.2. Structure

The reinforcement potential of any CB is governed, in some extent, by its structural level (aggregation degree). Herein, structure is determined by the size and shape of the aggregated primary particles [18]. A high structure in a CB indicates a high number of primary particles per aggregate, while low structure degree is related to a weak aggregation level [29]. It is important to highlight that different CBs could have similar particle size but no equal structures [56]. Higher structure is manifested in many ways in the rubber compound. For instance, higher viscosity, stiffness or elastic modulus, hardness, conductivity and easier dispersion, while low tear strength and swell [52, 54]. Figure 2 illustrates the structure of three different commercial CB where N990 exhibits low structure (low aggregation degree), whilst N121 shows a high one (high aggregation degree).

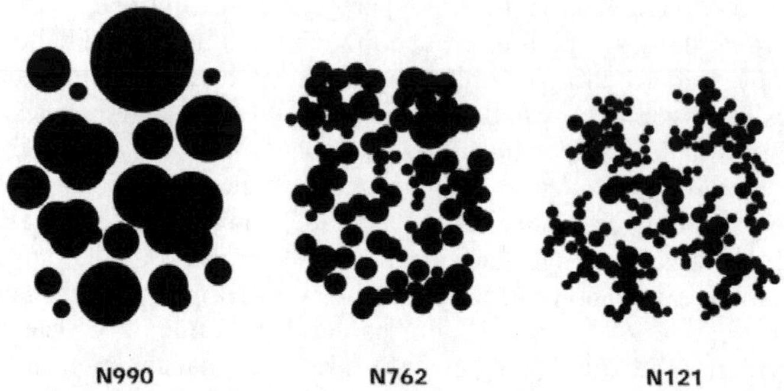

Figure 2. Structure representation of some CBs [25]

The CBp structure seems to be low since it commonly shows wider aggregates size distribution. In this way, Berki et al. [57] compared the CBp to the commercial CB N330 through SEM images, finding that CBp is less homogeneous than N330 and contains many large aggregates. This result is not surprising since, as mentioned previously, CBp is a mixture of different CBs. The CB structure is commonly determined by the *Oil Adsorption Analysis (OAN)*. Herein, the greater the volume of oil absorbed by the CB, the greater its structure [58]. The OAN value of commercial CB ranges from 58 to 105 $mL100g^{-1}$, while in the case of CBp this value has been found to be around 80 $mL100g^{-1}$ [9].

4.3. Surface area

Surface area is an important parameter for CB classification and also helps for predicting the rubber reinforcement potential. As mentioned before, surface area is related to particle size. In this sense, CBs with small particle size have more external surface ability for rubber interactions. In addition, CB pores also influence the surface area and promote the interaction between CB and rubber, although it depends on the pore opening size. Conventionally, pores are divided into three types according to size: (i) macropores (> 50 nm), (ii) mesopores (between 2 and 50 nm), and (iii) micropores (< 2 nm) [59]. San Miguel et al. [60] reported that the CBp obtained from the WTs pyrolysis at 700°C was limited in surface area and micropore volume but it exhibited a relatively large mesopores structure which can be helpful for large molecules size absorption. The common and accurate technique to determine the specific surface area is the N_2 adsorption known as Brunauer, Emmett, and Teller (BET), as described by the ASTM D6556. However, this analysis requires specific and qualified equipment, which makes it very expensive. For this reason, the CB industry commonly use cheaper techniques as the iodine adsorption (ASTM D1510) and the methylene blue adsorption (ASTM C1777). Typically, commercial CBs have surface areas (BET) between 40 and 155 m^2g^{-1} [59], whilst for CBp this property ranges between 57 and 89 m^2g^{-1}as reported in Table 2. It means that CBp surface area could be compared to the commercial CBs between references N774 and N330 (Table 1). BET surface area and iodine adsorption have a direct relation, although the proportionality constant depends on each CB type [61].

4.4. Surface chemistry

Surface chemistry in CB is usually related to oxygen surface functional groups (SFG), which improves the pigment wetting and dispersion, among others. Although it plays an important role in the polymer-filler interaction, no direct correlation has been demonstrated between the chemical groups present onto the CB surface and the related properties of rubber. Indeed, it has been stated that the filler polymer interaction in commercial CBs is mainly of physical nature (physisorption) [51]. For CBp, has been reported that functional groups can be obstructed by the inorganic matter, which have been promoted demineralization processes to remove these impurities to improve the filler-rubber interactions [13, 32, 48].

Generally speaking, some functional groups can appear at the edges of graphene layers usually formed by the presence of heteroatoms (H, O, N, S, halogens) which constitutes active sites [62]. These groups could have either acid or basic character. The acid groups are carboxylic acid, carboxylic anhydride, lactone, lactols, phenol, and hydroxyl. Whereas the basic character is related to the nitrogenous groups such as pyridine, nitrile, pyrrole, amine, amide, etc [62]. In general, high oxygen content is associated whit the presence of more acid groups while low oxygen content with basic groups [63].

There are different methods, either qualitative or quantitative, to characterize the SFG. For instance, infrared spectra is commonly used as a qualitative technique and the volumetric analysis, developed by P. H. Boehm (known as the Boehm method), is used as a quantitative tool. The latter method consists in a selective neutralization of the acid or basic SFG. The acid groups quantified by this method are carboxylic (acid and anhydride), lactonic (lactone and lactols) and phenolic including hydroxyls. Figure *3* shows a schematic representation of these SFG. Boehm method consider that NaOH neutralizes carboxylic, lactonic and phenolic; Na_2CO_3 reacts with carboxylic and lactonic; and $NaHCO_3$ reacts only with carboxylic groups. Regarding basic groups, they are quantified as total basic groups and are neutralizes by HCl [63–68].

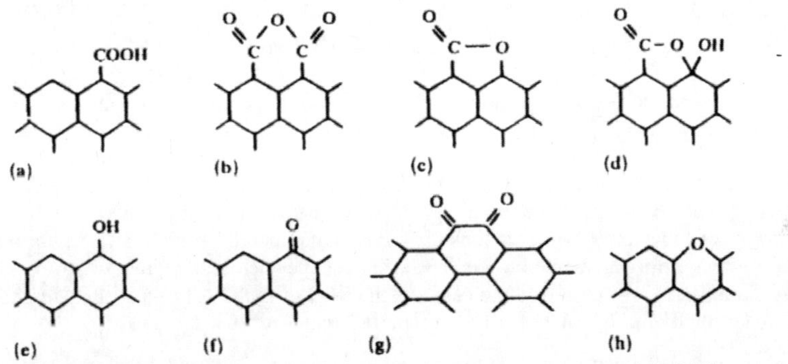

Figure 3. Surface functional groups of carbonaceous materials [63].
a) Carboxylic acid, b) Carboxylic anhydride, c) Lactone, d) Lactol, e) hydroxyl (phenol), f) Carbonyl, g) Quinone, h) substitution of carbon by oxygen (ether o xanthene)

On the other hand, in the infrared spectroscopy, carbonaceous materials can exhibit characteristic peaks around 3300-3600 cm^{-1} and 3100-3200 cm^{-1} which correspond to carboxylic acid vibration COOH and hydrogen bond due to the OH stretching. The peak near to 3400 cm^{-1} is characteristic from alcohol and phenol. Other important peaks are located around 1110, 1600 and 1500 cm^{-1} and are related to C-O vibration, carbonite bond (C=O) and the carbonyl-carbonate group, respectively [69]. In addition, is possible to find a peak near to 1200 cm^{-1}, which is characteristic of aromatic alcohols. Also, peaks around 2900 cm^{-1}, representing the symmetric and asymmetric C-H bond, indicating some residual organic matter onto the CBp surface [70]. Furthermore, peaks in the range of 1200-1300 cm^{-1} could also be found. They are linked to the vibration of C-O and C-OH molecules that represent phenolic or carboxylic compounds [71]. Table 4 shows the wavelength of some SFG commonly found in carbonaceous materials. The infrared spectra of CBp has been analysed by Li et al [70]. The authors reported that this CBp has similar absorption peaks as N330, but with weaker intensities for peaks at around 3435 and 1084 cm^{-1}, which are attributed to the stretching vibration of OH and CO, respectively.

Table 4. Infrared functional groups

Wavelength (cm^{-1})			Molecule or functional group	Ref.
3440			O-H de hydroxyl, phenol y/o carboxylic	[71]
3200-3640		1049-1276	Alcohols and absorbed water	[69, 72]
2500-3620		1160-1200	O-H de phenols	
2854 y 2920			-CH,-CH_2 y CH_3	
2600-3000			C-H	
2500-3300	1665-1760	1120-1200	Carboxylic acids	
	1740-1880	980-1300	Carboxylic anhydride	
	1675-1790	1160-1370	Lactones	[71]
	1585-1600		C=C aromatic	
	1590-1600	1100-1500	Carboxyl-carbonate	
		1000-1220	C-OH de phenols	
		1000-1300	C-O ether	[69, 71]
	1723-1735		C=O Carbonyl	[19, 71]
2250-2400			C=O	[72]
		1000-1200	C-O	

Throughout the years, efforts have been made to understand how the SFG reacts chemically with rubber molecules. However, no important results have been found. Instead, it is well-known that surface activity (related to surface chemistry) should be high in order to achieve a high interaction between the CB and the rubber matrix [73].

5. The potential used of CBp in reinforcing rubber

The incorporation of CB into rubber composites enhances the mechanical behaviour of the rubber good, from both physical and chemical interactions [59]. Physical interactions are related to the adsorption of free radicals coming from polymers onto the CB surface; while chemical interactions are associated with chemical reactions between these free radicals and the SFG of the CB. In this sense, that interaction is mainly related to the chemical surface, structure, particle size and surface area [74].

In general terms, the research works about the incorporation of CBp into styrene butadiene rubber (SBR) and/or natural rubber (NR), have concluded that commercial CBs (e.g. N660, N330, N550, N339, etc.) show higher reinforcing activity than CBp [57, 75–78]. Moreover, these research works have reported and compared the mechanical, dynamic mechanical thermal and the fracture mechanical behaviour when using different CBp in rubber formulations, as shown below.

5.1. Mechanical behaviour

The incorporation of CBp (with ~80 wt.% carbon content, ~20 wt.% ash and 75 m^2g^{-1} BET surface area) into SBR, resulted in the increase of both hardness (shore A) and tensile strength, as increasing the amount of CBp. The commercial reference N330 (80 m^2g^{-1} BET surface area -similar to CBp) has shown the same tendency but presented higher values of both properties as compared to CBp [57]. In addition, the same authors in other two papers, also found higher hardness and tensile strength in SBR formulations filled with N660 (BET of 35 m^2g^{-1}) at 60 phr than with CBp (the same used in ref. [57]) [75, 76]. This fact could be attributed to the inorganic compounds and carbonaceous deposits presented in CBp [75, 76].

Du et al [55] also evaluated the mechanical properties of SBR formulations with CBp (with 85.8 wt% of C and 13.3 wt.% of ash) and compared it to SBR filled with N330 and N774 at different loadings. They noticed that the increase of each filler content greatly improved the tensile strength, tear strength and modulus at 100% of strain, as compared to the unfilled rubber. For instance, when the CBp amount increased from 10 to 50 phr, the tensile strength of SBR composites increased by 9 times, while the tear strength increased by 3 times. However, the CBp led to a decrease of these properties, as compared to N774 and N330. Even though the authors stated that N774 can be partly replaced by CBp in some rubbers formulations.

The partial substitution of CB by CBp at a given amount, have resulted in different behaviours. For instance, Karabork et al [77] replaced partially the N550 at 50 phr by the CBp produced at 500-550°C in a fixed bed reactor (FBR) and evaluated its performance in NR-SBR blends. They found a decrease in hardness (shore A), tear strength and the tensile strength, as increasing CBp amount. On the other hand, Berki et al [57] found a decrease in the tensile strength and hardness when increase the amount of CBp in the partial substitution of N330 by CBp at different loads (30, 45 and 60 phr), although the tear strength increased. Likewise, Cataldo [78] evaluated the partial substitution of N339 (90 m^2g^{-1} BET surface area) by a CBp (produced at 700°C with 81 m^2g^{-1} BET surface area and about 10 wt.% ash) into NR-SBR composites. Similar to Karabork et al [77] and Berki et al [57], the author also observed a decrease of tensile strength and hardness IRHD with the increase of CBp content in the blends.

Another important property affected by the incorporation of CBp is the tensile strain at break. By increasing the filler amount, either CB or CBp as well as combinations of both, the tensile strain decreases. That phenomenon occurred because during filler dispersion in rubber, part of the elastomer fills the void volume of the CB aggregates. The elastomer portion located within the irregular contours of the aggregates is unable to participate fully in the deformation. If the CB presents high structure, the elastomer filled into the filler void volume shows lower deformation which means low tensile strain value [59]. Tensile strain was lower in N330 – SBR composites than composites filled with CBp [57]. SBR composites filled with N660 showed also lower tensile strain than those filled with CBp [75, 76].

Due to the low performance of CBp in rubber composites, different treatments have been tested to improve the CBp quality such as acid treatments and the use of coupling agents [12, 13]. For instance, Zhou et al [12] evaluated a washing process of CBp with HNO_3 (5wt.%) and the use of titanate coupling reagent into the washed CBp. Those modifications resulted in an increase of the tensile strength compared to the untreated CBp.

5.2. Dynamic mechanical thermal behaviour
The dynamic mechanical thermal analysis (DMTA) leads to obtain the storage modulus (E') and the mechanical loss factor (tan δ). The maximum tan δ peak is found in a marginal change in glass transition temperature (T_g) [57]. In general terms, the tan δ at 0°C is related to the tyre performance, high value is associated with good skid resistance in wet or cold road conditions and low tan δ value at 60°C is correlating with low rolling resistance. However, it is difficult to obtain both high tan δ at 0° and low tan δ at 60°C at the same time, therefore, a balance between those values may suggest a suitable rubber compound [79]. Furthermore, the DMTA also reports the E', specifically the rubber and glassy modules which are read at $T=T_g+30°C$ and $T=T_g-30°C$, respectively. Both modules increase as increasing filler content, while a decrease in tan δ is shown as increase filler amount as well [57, 75, 76]. That decrease in tan δ is associated with the increasing of filler-rubber interaction (reinforcing effectiveness) due to the reduction of rubber chains mobility in the filler-rubber interphase caused by the increase of filler amount [77]. In addition, the rubber modulus is used to calculate the apparent crosslink density (M_c) (also known as apparent mean molecular weight). Although the M_c is an apparent value because it is affected not only by the filled rubber crosslinking but also by the rubber-filler and filler-filler interactions, lower M_c values indicates high crosslink [57, 75].

In this sense, SBR-CBp composites (mentioned in section 5.1. ref. [57]) showed an increase of both rubber and glassy modules, a decrease of tan δ and a decrease of M_c by increasing the CBp content. In all filler loadings (30, 45 and 60 phr), the M_c and tan δ properties were lower, while the rubber and glassy modules were higher in the N330 composites compared to the CBp ones [57]. In the case of the comparison of the CBp with the N660, which is a lower reinforcing grade CB respect to N330, the SBR filled at 60 phr showed similar M_c and rubber modulus, while the glassy modulus was higher and the tan δ was lower in the N660 composites compared to the CBp ones [75].

As well as the mechanical behaviour, in the partial substitution of CB by CBp, Karabork et al found in the composites mentioned before in section 5.1. that the tan δ first increased and then decreased by increasing the CBp amount. However, all samples with CBp showed higher tan δ than the N550 composites due to CBp caused a decrease in the elastic behaviour which is reflected in the greater tan δ value. Respect to the E', the same samples showed a decrease in this property by the CBp increase, as expected. This fact means a weaker reinforcing effect of CBp than the commercial N550 which is attributed to the difference in morphology, structure and surface area between those two fillers [77].

Another important parameter from the DMTA is the Payne effect which reflects the agglomeration tendency or the filler reinforcing activity. The higher this value, the more active the filler is. The Payne effect is read in the plot of E' as a function of the strain and is measured by the difference in the E' modulus at 0.01 and 10% tensile strain [57, 75]. The SBR-CBp composites of ref. [57] (mentioned in section 5.1.), showed grater Payne effect value by the increasing CBp filled, but it was lower than the CB N330 at all loadings (30, 45 and 60 phr) [57]. Contrary, the SBR composites filled at 60 phr of CBp (the same used in ref. [57]) was higher than the ones filled with CB N660 at 60 phr as well. Furthermore, the combinates filler (30 phr CBp and 30 phr N660), yields the same Payne effect that the ones of N660 alone [75].

5.3. Fracture mechanical behaviour

The fracture mechanical test is used to investigate the crack propagation and the failure process. Generally, it is applied the crack tip opening displacement (CTOD) to study the crack propagation by analysing a sequence of images to detect the point where fracture started to propagate. This test leads to calculate the energy dissipated during fracture per unit of newly created fracture surface area (J-integral). The critical value of J-integral (J_c) is found on the crack onset. Due to it is hardly to find optically, some authors settle the J_c value at CTOD of 0.5 mm [75, 76] as well as 0.1 mm [57].

The SBR with CBp composites (mentioned in section 5.1. ref. [57]) showed slightly higher Jc values compared to SBR with N330 [57]. However, the SBR with CBp (the same used in ref. [57]) at 60 phr filling showed the same value of J_c at CTOD of 0.5 mm to SBR with N660 at 60 phr filling as well. The difference between CBp and N330 is attributed to the dispersion state. The wide range of particle sizes as well as the large particle agglomerates of the CBp hinder crack propagation [57].

6. Conclusions

This work considered the literature review not only of the CBp properties but also its potential as a reinforcing-filler in rubber applications. The main differences between CBp and CB is the inorganic content, the carbonaceous deposits on the CBp surface and the heterogeneous morphology. These factors affect structure, surface area and particle size, among others, and are the main responsible for the mechanical properties detrimental when CBp is used in rubber formulations. Hence, some post treatments to the CBp can overcome these challenges. For instance, a demineralization process can remove the inorganic compounds as well as the carbonaceous deposits. A thermal treatment is also useful to remove residual volatile matter. Other treatment is a milling process to reduce and homogenize the aggregate size of CBp. Those post treatments have been suggested in order to improve the CBp quality and enhance its performance in rubber composites. Despite of the weak performance of rubber formulations using CBp reported in literature, it is possible its partial substitution in rubber goods that not require high mechanical properties.

On the other hand, CBp quality seems to depend on both pyrolysis conditions (temperature, pressure, particle size, heating rate, residence time (solid and vapours)) and pyrolysis reactor technology (rotary kiln, fixed bed, fluidized bed and auger). All the parameters influence the CBp properties and hence, their performance in rubber formulations. Also, tyre type seems to affect the pyrolysis products yields and properties. In addition, a proper selection of waste tyres can enhance the CBp properties, as the CBp is more homogeneous as compared with that obtained when different types of waste tyres are pyrolyzed.

Acknowledgments
The authors would like to express all their gratitude to COLCIENCIAS for the financial support given by the research project (FP44842-026-2016-2002). N. Cardona-Uribe is in debt with the Integrated Center for the Development of Research (CIDI) at Pontificia Bolivariana University and its program "Formación Investigativa" for the MSc scholarship.

References
[1] Machin EB, Pedroso DT, and de Carvalho Jr. JA. Energetic valorization of waste tires. 2017, *Renew. Sustain. Energy Rev.*, **68**, **Part 1**, pp. 306–315.
[2] Barroso FJ. Un gran incendio en Seseña arrasa el mayor cementerio de neumáticos de España Un gran incendio en Seseña arrasa el mayor cementerio de neumáticos de España *El país* 2016 https://goo.gl/DDEfZi (accessed 8 June 2017).
[3] Orion engineered carbon. Manufacturing processes Manufacturing processes 2013 https://goo.gl/VBc8kH (accessed 15 June 2017).
[4] Martínez JD, Murillo R, García T, et al. Demonstration of the waste tire pyrolysis process on pilot scale in a continuous auger reactor. 2013, *J. Hazard. Mater.*, **261**, pp. 637–645.
[5] Aylón E, Fernández-Colino A, Murillo R, et al. Valorisation of waste tyre by pyrolysis in a moving bed reactor. 2010, *Waste Manag.*, **30**, pp. 1220–1224.
[6] Murillo R, Aylón E, Navarro MV, et al. The application of thermal processes to valorise waste tyre. 2006, *Fuel Process. Technol.*, **87**, pp. 143–147.
[7] Williams PT. Pyrolysis of waste tyres: a review. 2013, *Waste Manag.*, **33**, pp. 1714–28.
[8] Martínez JD, Murillo R, and García T. Production of carbon black from the waste tires pyrolysis. 2013, *Bol. Grup. Español Carbón*, pp. 10–14.
[9] Norris CJ, Hale M, and Bennett M. Pyrolytic carbon: factors controlling in-rubber performance. 2014, *Plast. Rubber Compos.*, **43**, pp. 245–256.
[10] Mui ELK, Ko DCK, and McKay G. Production of active carbons from waste tyres—a review. 2004, *Carbon*, **42**, pp. 2789–2805.
[11] Aylón E. *Diseño y optimización de una instalación para el tratamiento de neumáticos fuera de uso* 2007 PhD thesis. Universidad de Zaragoza].
[12] Zhou J, Yu T, Wu S, et al. Inverse Gas Chromatography Investigation of Rubber Reinforcement by Modified Pyrolytic Carbon Black from Scrap Tires. 2010, *Ind. Eng. Chem. Res.*, **49**, pp. 1691–1696.
[13] Zhou J, Wu S, Yu T, et al. Modified pyrolytic carbon black from scrap tires and its reinforcement performance in natural rubber. 2011, *Proc. - Int. Conf. Comput. Distrib. Control Intell. Environ. Monit. CDCIEM 2011*, pp. 472–475.
[14] Goulaouic S, Foucaud L, Bennasroune A, et al. Identification of polycyclic aromatic hydrocarbons in carbon black with reference to cancerogenic risk in tire production. 1979, *Am. Ind. Hyg. Assoc. J.*, **40**, pp. 644–652.
[15] Locati G, Fantuzzi A, Consonni G, et al. Effect of Polycyclic Aromatic Hydrocarbons and Carbon Black Particles on Pro-Inflammatory Cytokine Secretion: Impact of PAH Coating Onto Particles. 2008, *J. Immunotoxicol.*, **5**, pp. 337–345.
[16] Rodat S, Abanades S, and Flamant G. Co-production of hydrogen and carbon black from solar thermal methane splitting in a tubular reactor prototype. 2010, *Sol. Energy*, **85**, pp. 645–652.
[17] Lefton R. Super Pollutants 101: How To Cut Climate Impacts And Safeguard Public Health

Super Pollutants 101: How To Cut Climate Impacts And Safeguard Public Health *ThinkProgress* 2013 https://goo.gl/h5kmBo (accessed 5 June 2017).
[18] IARC. Carbon black. 2010, *Monogr Eval Carcinog Risks Hum*, **93**, pp. 43–191.
[19] Betancur M. *Obtención de carbón activado a partir de neumáticos usados mediante la incorporación de calcio y su aplicación en proceso de adsorción* 2010 PhD thesis. Ingeniería Química. Universidad Pontificia Bolivariana.
[20] ICBA. Carbon black user's guide Carbon black user's guide *International Carbon Black association* 2016 https://goo.gl/fxAZfe (accessed 17 March 2016).
[21] Cabot. Negros de humo para refuerzo del elastómero Negros de humo para refuerzo del elastómero 1995 http://www.cabotcorp.com.co (accessed 17 March 2016).
[22] EPA. Carbon Black Carbon Black *Environmental Protection Agency* 2015 https://goo.gl/2u3vUC (accessed 15 June 2017).
[23] Studebaker ML. The Chemistry of Carbon Black and Reinforcement. 1957, *Rubber Chem. Technol.*, **30**, pp. 1400–1483.
[24] Cancarb. What is thermal carbon black What is thermal carbon black 2017 https://goo.gl/vkTtJD (accessed 16 June 2017).
[25] Cancarb. Physical & Chemical Properties Physical & Chemical Properties 2016 https://goo.gl/SuZm2o (accessed 10 May 2016).
[26] Xuzhou First Chemical Co. Ltd. Carbon Black N220/N330/N550/N660 Carbon Black N220/N330/N550/N660 https://goo.gl/uf9oiC (accessed 12 December 2016).
[27] Martínez JD, Puy N, Murillo R, et al. Waste tyre pyrolysis – A review. 2013, *Renew. Sustain. Energy Rev.*, **23**, pp. 179–213.
[28] Darmstadt H, Roy C, and Kaliaguine S. Characterization of pyrolytic carbon blacks from commercial tire pyrolysis plants. 1995, *Carbon*, **33**, pp. 1449–1455.
[29] Martínez Ángel JD. *Waste tire valorization by pyrolysis: production and application of the products* 2014 PhD thesis. Universidad de Zaragoza pp.239.
[30] Darmstadt H, Roy C, and Kaliaguine S. Esca Characterization of Commercial Carbon-Blacks and of Carbon-Blacks from Vacuum Pyrolysis of Used Tires. 1994, *Carbon*, **32**, pp. 1399–1406.
[31] Huang K, Gao Q, Tang L, et al. A comparison of surface morphology and chemistry of pyrolytic carbon blacks with commercial carbon blacks. 2005, *Powder Technol.*, **160**, pp. 190–193.
[32] Henry KL. *Upgrading pyrolytic tyre char through acid-alkali demineralisation* 2015 MSc thesis. Chemical Engineering. Stellenbosch University pp.186.
[33] Quek A, and Balasubramanian R. Preparation and characterization of low energy post-pyrolysis oxygenated tire char. 2011, *Chem. Eng. J.*, **170**, pp. 194–201.
[34] Díez C, Martínez O, Calvo LF, et al. Pyrolysis of tyres. Influence of the final temperature of the process on emissions and the calorific value of the products recovered. 2004, *Waste Manag.*, **24**, pp. 463–9.
[35] Kyari M, Cunliffe A, and Williams PT. Characterization of oils, gases, and char in relation to the pyrolysis of different brands of scrap automotive tires. 2005, *Energy & Fuels*, **19**, pp. 1165–1173.
[36] González JF, Encinar JM, Canito JL, et al. Pyrolysis of automobile tyre waste. Influence of operating variables and kinetics study. 2001, *J. Anal. Appl. Pyrolysis*, **58–59**, pp. 667–683.
[37] Li SQ, Yao Q, Chi Y, et al. Pilot-scale pyrolysis of scrap tires in a continuous rotary kiln reactor. 2004, *Ind. Eng. Chem. Res.*, **43**, pp. 5133–5145.
[38] Zhang X, Wang T, Ma L, et al. Vacuum pyrolysis of waste tires with basic additives. 2008, *Waste Manag.*, **28**, pp. 2301–10.
[39] López FA, Centeno TA, Alguacil FJ, et al. Distillation of granulated scrap tires in a pilot plant. 2011, *J. Hazard. Mater.*, **190**, pp. 285–92.
[40] Ucar S, Karagoz S, Ozkan AR, et al. Evaluation of two different scrap tires as hydrocarbon source by pyrolysis. 2005, *Fuel*, **84**, pp. 1884–1892.
[41] Galvagno S, Casu S, Casabianca T, et al. Pyrolysis process for the treatment of scrap tyres:

preliminary experimental results. 2002, *Waste Manag.*, **22**, pp. 917–923.
[42] Napoli A, Soudais Y, Lecomte D, et al. Scrap tyre pyrolysis: Are the effluents valuable products? 1997, *J. Anal. Appl. Pyrolysis*, **40–41**, pp. 373–382.
[43] Berrueco C, Esperanza E, Mastral FJ, et al. Pyrolysis of waste tyres in an atmospheric static-bed batch reactor: Analysis of the gases obtained. 2005, *J. Anal. Appl. Pyrolysis*, **74**, pp. 245–253.
[44] Lopez G, Olazar M, Amutio M, et al. Influence of tire formulation on the products of continuous pyrolysis in a conical spouted bed reactor. 2009, *Energy & Fuels*, **23**, pp. 5423–5431.
[45] Aylón E, Fernández-Colino A, Navarro M V., et al. Waste tire pyrolysis: Comparison between fixed bed reactor and moving bed reactor. 2008, *Ind. Eng. Chem. Res.*, **47**, pp. 4029–4033.
[46] Chaala A, Roy C, and Ait-Kadi A. Rheological properties of bitumen modified with pyrolytic carbon black. 1996, *Fuel*, **75**, pp. 1575–1583.
[47] Aranda A, Murillo R, García T, et al. Steam activation of tyre pyrolytic carbon black: Kinetic study in a thermobalance. 2007, *Chem. Eng. J.*, **126**, pp. 79–85.
[48] Chaala A, Darmstadt H, and Roy C. Acid-base method for the demineralization of pyrolytic carbon black. 1996, *Fuel Process. Technol.*, **46**, pp. 1–15.
[49] Shen B, Lu F, and Liu T. Modification of carbon black produced from pyrolysis of scrap tires. 2009, *Huagong Xuebao/CIESC J.*, **60**, pp. 2327–2331.
[50] Burg P, and Cagniant D. *Chemistry and Physics of Carbon* 2008 [New York: CRC Press Taylor & Francis Group] pp.129-175.
[51] Fröhlich J, Niedermeier W, and Luginsland HD. The effect of filler-filler and filler-elastomer interaction on rubber reinforcement. 2005, *Compos. Part A Appl. Sci. Manuf.*, **36**, pp. 449–460.
[52] Birla Carbon. Fundamental properties of carbon black Fundamental properties of carbon black 2017 https://birlacarbon.com/whats-trending/carbon-black/ (accessed 7 May 2017).
[53] Wiegand WB, and Synder JW. Some Properties of Carbon Black IV Color. 1934, *Ind. Eng. Chem.*, **26**, pp. 413–419.
[54] Norman DT. Rubber grade carbon blacks Rubber grade carbon blacks 2014 https://goo.gl/YWeypr (accessed 20 March 2017).
[55] Du A, Wu M, Su C, et al. The characterization of Pyrolytic carbon black prepared from used tires and its application in Styrene–Butadiene Rubber (SBR). 2008, *J. Macromol. Sci. part B Phys.*, **47**, pp. 268–275.
[56] Medalia AI, and Heckman FA. Morphology of aggregates—II. Size and shape factors of carbon black aggregates from electron microscopy. 1969, *Carbon*, **7**, pp. 567–582.
[57] Berki P, Göbl R, and Karger-Kocsis J. Structure and properties of styrene-butadiene rubber (SBR) with pyrolytic and industrial carbon black. 2017, *Polym. Test.*, **61**, pp. 404–415.
[58] Lin C, and Chung DDL. Effect of carbon black structure on the effectiveness of carbon black thermal interface pastes. 2007, *Carbon*, **45**, pp. 2922–2931.
[59] Spahr ME, and Rothon R. *Polymers and Polymeric Composites: A Reference Series* 2016 [Berlin, Heidelberg: Springer] pp.1-31.
[60] San Miguel G, Fowler GD, and Sollars CJ. Pyrolysis of tire rubber: porosity and adsorption characteristics of the pyrolytic chars. 1998, *Ind. Eng. Chem. Res.*, **37**, pp. 2430–2435.
[61] Puri BR, and Bansal RC. Iodine adsorption method for measuring surface area of carbon blacks. 1965, *Carbon*, **3**, pp. 227–230.
[62] Maldonado-Hórdar FJ, Morales-Torres S, Pérez-Cardenas AF, et al. Química superficial de los materiales de carbón. 2011, *Bol. Grup. Español Carbón*, **20**, pp. 10–15.
[63] Boehm HP. Some aspects of the surface chemistry of carbon blacks and other carbons. 1994, *Carbon*, **32**, pp. 759–769.
[64] Ciobanu M, Lepadatu A-M, and Asaftei S. Chemical and electrochemical studies of carbon black surface by treatment with ozone and nitrogen oxide. 2016, *Mater. Today Proc.*, **3**, pp. 252–257.
[65] Panchompoo J, Aldous L, and Compton RG. Size-effects in the chemical modification of carbon black nanoparticles with 4-nitroaniline. 2010, *New J. Chem.*, **34**, p. 2643.
[66] Luo J, Lu J, Niu Q, et al. Preparation and characterization of benzoic acid-modified activated

carbon for removal of gaseous mercury chloride. 2015, *Fuel*, **160**, pp. 440–445.
[67] Matsumura Y, Hagiwara S, and Takahashi H. Automatic potentiometric titration of surface acidity of carbon black. 1976, *Carbon*, **14**, pp. 163–167.
[68] Boehm HP. Chemical Identification of Surface Groups. 1966, *Adv. Catal.*, **16**, pp. 179–274.
[69] Fanning PE, and Vannice MA. A DRIFTS study of the formation of surface groups on carbon by oxidation. 1993, *Carbon*, **31**, pp. 721–730.
[70] Li S, Wan C, Wu X, et al. Core-shell structured carbon nanoparticles derived from light pyrolysis of waste tires. 2016, *Polym. Degrad. Stab.*, **129**, pp. 192–198.
[71] Khelifi A, Almazán-Almazán MC, Pérez-Mendoza M, et al. Influence of nitric acid concentration on the characteristics of active carbons obtained from a mineral coal. 2010, *Fuel Process. Technol.*, **91**, pp. 1338–1344.
[72] Chen JP, and Wu S. Acid/Base-treated activated carbons: Characterization of functional groups and metal adsorptive properties. 2004, *Langmuir*, **20**, pp. 2233–2242.
[73] Darmstadt H, Cao N-Z, Pantea DM, et al. Surface activity and chemistry of thermal carbon blacks. 2000, *Rubber Chem. Technol.*, **73**, pp. 293–309.
[74] Ma J-H, Zhang L-Q, and Wu Y-P. Characterization of filler-rubber interaction, filler network structure, and their effects on viscoelasticity for styrene-butadiene rubber filled with different fillers. 2013, *J. Macromol. Sci. Part B*, **52**, pp. 1128–1141.
[75] Berki P, and Karger-Kocsis J. Comparative Properties of Styrene-Butadiene Rubbers (SBR) Containing Pyrolytic Carbon Black, Conventional Carbon Black, and Organoclay. *Journal of Macromolecular Science, Part B: Physics*; 55.
[76] Berki P, and Karger-Kocsis J. Tensile and fracture mechanical properties of styrene-butadiene rubbers (SBR) filled with industrial and pyrolytic carbon blacks, and organoclay. 2017, *Mater. Sci. Forum*, **885**, pp. 292–297.
[77] Karabork F, and Tipirdamaz ST. Influence of pyrolytic carbon black and pyrolytic oil made from used tires on the curing and (dynamic) mechanical properties of natural rubber (NR)/styrene-butadiene rubber (SBR) blends. 2016, *Express Polym. Lett.*, **10**, pp. 72–82.
[78] Cataldo F. Preparation of pyrolytic carbon black from scrap tire rubber crumb and evaluation in new rubber compounds. 2005, *Macromol. Mater. Eng.*, **290**, pp. 463–467.
[79] Ko JY, Prakashan K, and Kim JK. New silane coupling agents for silica tire tread compounds. 2012, *J. Elastomers Plast.*, **44**, pp. 549–562.

IOP Conf. Series: Materials Science and Engineering **437** (2018) 012013 doi:10.1088/1757-899X/437/1/012013

Practical PID controller implementation for the speed control of a motor generator system

M Paez[1]**, J Viola**[1] **and L Angel**[1]

[1] Departament of Electronic Engineering, Universidad Pontificia Bolivariana, Bucaramanga Colombia

jairo.viola@ upb.edu.co

Abstract. Windup effect in a PID controller is the condition of cumulative error in the integral action. It saturates the system actuators, letting the system out of control with a higher controller action. Putting a saturator for the controller action, it is limited into a linear range of operating values. However, it does not eliminate the cumulative error of the integral term. This paper presents the speed control of a motor generator system using a digital PID controller with antiwindup scheme to eliminate the cumulative error of the integral term of controller action. Initially, a linear model of the motor generator system was identified. From the linear model of the system, a PID controller was designed using pole placement technique. Controller discrete implementation was perform using the backward difference discretization methodology. PID controller consider the antiwindup scheme. PID controller practical implementation employed Matlab Stateflow and a data acquisition card. Obtained results showed that discrete implementation of the PID controller has a good performance against external disturbances and random noise.

1. Introduction

PID controller is one of the most employed control techniques to control industrial process due to its easy design and implementation [1,2]. The design of PID controllers is to find the value of the proportional, integral, and derivate parameters to accomplish the desired closed loop specifications. The wind-up effect of the integral action is a common problem in the PID controllers. It is caused by the cumulative error on the integral term which can lead to the actuator saturation, causing the system to behave as if it were in open loop. This is a not desirable condition in the system because it disappear the control effect over the system.

To solve the wind-up effect, there are many antiwind-up methods as incremental algorithm, back calculation, conditional integration and the observer approach [1–5]. These techniques act over the controller integral action when the actuator reach the saturation limits.

This paper presents the design and implementation of a speed control for a motor-generator system employing a digital PID controller tuned by the pole placement technique. PID controller is discretized employing the backward difference approximation with the back calculation antiwind-up technique. Initially, the PID controller structure and the back calculation antiwind-up technique are presented. Then, the discrete PID controller model is obtained with the backward difference method. After that, the motor generator system is descripted and a linearized model of the system is identified. From the linearized model, PID controller is designed using the pole placement method. Then, the digital PID controller and the antiwind-up technique are implemented employing the Matlab stateflow toolbox. To evaluate the performance of the digital PID controller, it is tested in the presence of external disturbances, random noise in the feedback loop and a resistive payload in the generator output. As performance indexes, the root median square error (RMSE) is employed for the system time response and the root mean square value (RMS) is employed for the control action. This paper is structured as follows. First, the PID controller structure and the antiwind-up technique are explained. Second, the PID controller discretization is developed. Third, the motor generator system description and identification are shown. Fourth, the practical implementation of the digital PID controller and its validation against external disturbances are performed. Finally, the conclusions are presented.

2. PID controller

From [1], a PID controller is represented in the frequency domain by the transfer function (1), where $U(s)$ is the control action, $E(s)$ the feedback system error, k_p the proportional constant of the system, τ_i the integral time constant and t_d the derivate time constant.

$$C(s) = \frac{U(s)}{E(s)} = k_p\left(1 + \frac{1}{\tau_i s} + \tau_d s\right) \tag{1}$$

3. Antiwind-up back calculation method

Figure 1 shows a block diagram representation of the back calculation antiwind-up technique with a PID controller. As shown, the system has an additional path, generated by measuring the actual actuator output or the output of a mathematical model of the actuator with saturation, forming a new error signal (es), which represent the difference between the controller output (v) and the actuator output (u). This signal is feedback to the integrator via gain $1/T_t$. This signal is zero when the actuator is not saturated and will have no effect on the control action. When the actuator becomes saturated, the signal (s) is different from zero, so, the controller output is placed at the upper or lower saturation limit and the integral term of the controller is recalculated so that its new value changes dynamically using the time constant T_t. This constant must be greater than τ_d and smaller than τ_i. A heuristic rule suggested by [1] stablish that $T_t = \sqrt{\tau_i \tau_d}$.

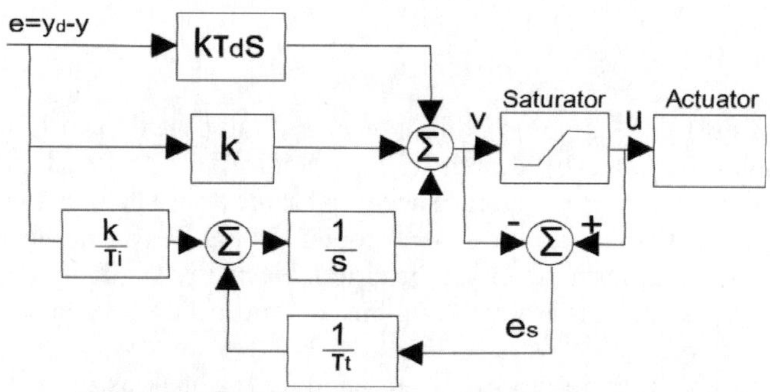

Figure 1. Back calculation antiwind-up scheme

4. PID controller discretization

The PID controller (1) can be discretized by different discretization techniques [6]. In this paper, the backward difference technique is employed, which approach the derivate of $x(t)$ into a differences equation as shown in (2), where T_s is the sampling time.

$$\frac{dx(t)}{dt} \approx \frac{x(k) - x(k-1)}{T_s} \tag{2}$$

IOP Conf. Series: Materials Science and Engineering **437** (2018) 012013 doi:10.1088/1757-899X/437/1/012013

From (2), the relation between the Laplace transform and Z transform on the frequency domain is given by (3).

$$s = \frac{1 - z^{-1}}{T_s} \tag{3}$$

Replacing (3) in (1)

$$C(z) = k_p + \frac{k_p T_s}{\tau_i}\left(\frac{1}{1 - z^{-1}}\right) + \frac{k_p \tau_d (1 - z^{-1})}{T_s} \tag{4}$$

Since $C(z) = U(z)/E(z)$, (4) can be expressed as:

$$U(z) = k_p E(z) + \frac{k_p T_s}{\tau_i}\left(\frac{1}{1 - z^{-1}}\right) E(z) + \frac{k_p \tau_d (1 - z^{-1})}{T_s} E(z) \tag{5}$$

$U(z)$ can be represented in function of the proportional $U_p(z)$, integral $U_i(z)$ and derivate $U_d(z)$ components as shown in (6).

$$U(z) = U_p(z) + U_i(z) + U_d(z) \tag{6}$$

where:

$$\begin{aligned} U_p(z) &= k_p E(z) \\ U_i(z) &= \frac{k_p T_s}{\tau_i}\left(\frac{1}{1 - z^{-1}}\right) \\ U_d(z) &= \frac{k_p \tau_d (1 - z^{-1})}{T_s} E(z) \end{aligned} \tag{7}$$

The representation given by (6) and (7) allows to implement the antiwind-up technique later. From (6) and (7), the following differences equations are obtained which represents the discrete PID controller model.

$$U(k) = U_p(k) + U_i(k) + U_d(k) \tag{8}$$

where:

$$\begin{aligned} U_p(k) &= k_p e(k) \\ U_i(k) &= U_i(k-1) + \frac{k_p T_s}{\tau_i} e(k) \\ U_d(k) &= \frac{k_p \tau_d (e(k) - e(k-1))}{T_s} \end{aligned} \tag{9}$$

5. Motor generator system

Fig 4 shows the motor generator system. This system is a motor-generator conformed by two DC motors Bodine brand with a maximum speed of 2500 RPM and $\pm 130v$ DC nominal voltage. As can be shown, the motors M1 and M2 are mechanically coupled by its axis where M1 works as motor and M2 as generator. M1 has a power driver M4 with an $\pm 10v$ input. A tachogenerator M3 with $\pm 7v$ output voltage is coupled to M2 for the speed feedback of the system.

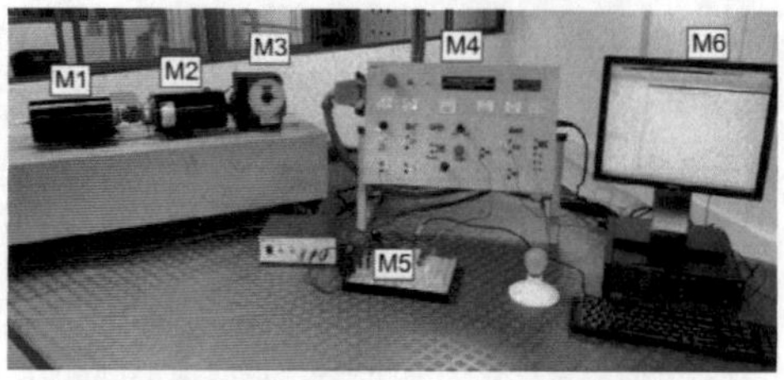

Figure 2. Motor Generator System

5.1. System identification

For the motor-generator system identification, the system is excited with a stepped speed profile shown in Figure 3 This profile allows to observe the dynamical behavior of the system for different operating points in both turn senses. As output signal for the identification process the tachogenerator voltage M3 is employed.

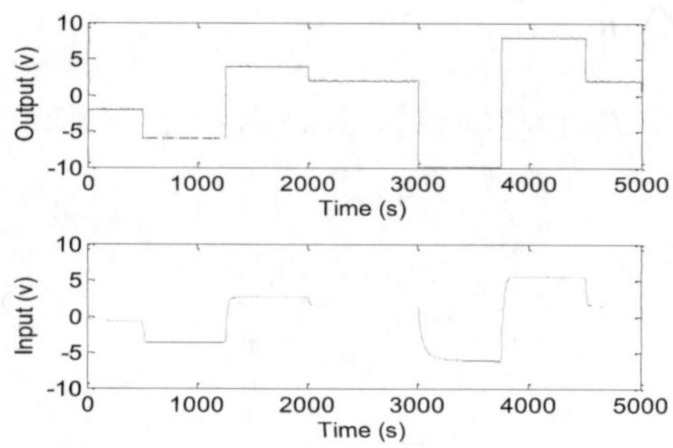

Figure 3. Identificacion trajectory

Data acquisition of the system is performed using a National Instruments data acquisition card NI DAQ 6008 M5 altogether with Matlab Simulink M6, where the steeped speed profile is generated and the system output is captured. From the acquired data, the model of the system is identified using the Matlab Identification Toolbox, resulting in a second order system with real poles given by (10) with a fit of 93%.

$$G_p(s) = \frac{0.62}{(21.725s + 1)(5.12s + 1)} \tag{10}$$

5.2. PID controller design

From the model of the system (10), the PID controller is tuned by the pole placement method. It is desired a closed loop overshoot of 10% and a setting time of $150s$. For this, a damping ratio $\zeta = 0.6$ and a frequency of $w_n = 0.04\ rad/s$. For a second order system with real poles (11) whose structure coincide with (10), the constants of the PID controller are calculated using the expressions (12) given by [1], where τ_1 and τ_2 are the time constants of the system k is the system gain and α is the location of a non-dominant pole of the closed loop system. Using (12) with $\alpha = 10$, the obtained constants for the PID controller are $k_p = 2.98$, $\tau_i = 19.02$ y $\tau_d = 15.35$.

$$G_p(s) = \frac{k}{(\tau_1 s + 1)(\tau_2 s + 1)} \tag{11}$$

$$k = \frac{\tau_1\tau_2 w_n^2(1+2\alpha\zeta)-1}{k} \quad \tau_i = \frac{\tau_1\tau_2 w_n^2(1+2\alpha\zeta)-1}{\tau_1\tau_2\alpha w_n^3} \quad \tau_d = \frac{\tau_1\tau_2 w_n(\alpha+2\zeta)-\tau_1-\tau_2}{\tau_1\tau_2 w_n^2(1+2\alpha\zeta)-1} \tag{12}$$

5.3. PID controller digital implementation using Matlab Stateflow

The digital implementation of the PID controller uses the difference equations (9) with the constants calculated above and a sampling time $T_s = 0.1\ s$. For the digital implementation of the back calculation antiwind-up technique proposed in Section 3, the pseudocode presented in Table1 is employed, where uk is the total control action, uik is the integral control action, lim_sat_p and $\lim\_sat\_n$ are the actuator saturation limits and T_t is the dynamic time reset constant of the integral action.

Table 1. Pseudocode for antiwindup back calculation digital implementation

Read *uk, uik, lim_sat_p, lim_sat_n*

If *uk>lim_sat_p* ***do:***
$uik = uik + (1/T_t)(lim\_sat\_p - uk)$
$uk = lim\_sat\_p$

else if $uk < lim\_sat\_n$ ***do:***
$uik = uik + (1/T_t)(lim\_sat\_n - uk)$
$uk = lim\_sat\_n$

else do:
$uk = uk$

To implement the digital PID controller tuned by the pole placement technique the matlab stateflow toolbox is employed with the data adquisition card NI DAQ which close the control loop for the motor generator system. Figure 4 shows the implementation of the PID controller employing the matlab stateflow toolbox. As can be observed, the digital implementation uses two states, the start state, and the control state. The start state is the first to be executed and it initializes the constants of the PID controller. In the control state, the error signal and the individual control actions are calculated as well as the antiwind-up calculation and the total control action. Variables uik_1, udk_1 and ek_1 are the previous values of the integral action, the derivate action, and the error signal.

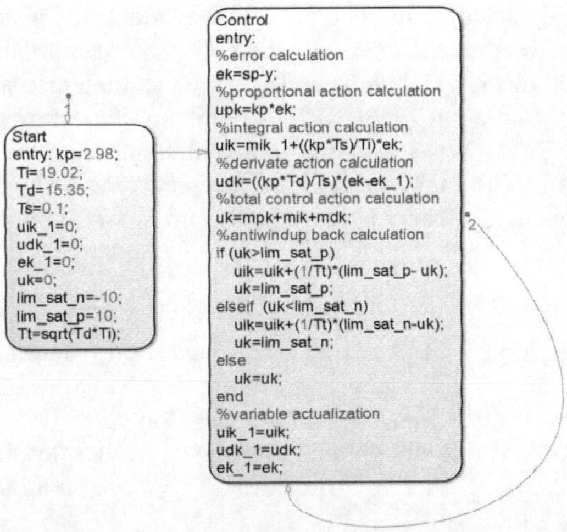

Figure 4. Matlab Stateflow PID implementation with antiwind-up back calculation technique

Figure 5 shows the motor-generator system time response and the control action of the digital PID controller tuned by the pole placement method for a stepped path which belongs to different desired speeds in both turn senses. As can be observed, The PID controller has a good tracking response and the control action is between the saturation limits.

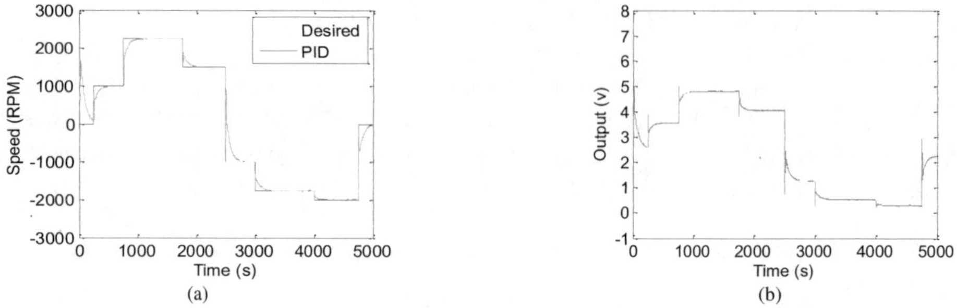

Figure 5. Time response (a) and control action (b) for the motor generator system with the digital PID controller

6. Robustness analysis

To evaluate the robustness of the digital PID controller, the time response of the system and the control action of the PID controller are tested in presence of external disturbance, random noise in the feedback loop and resistive payload at the generator output. As performance indexes, the RMSE value is selected for the time response and the RMS value is selected for the control action. Table 2 shows the obtained performance index values in presence of different disturbances. As shown, the PID controller performance is affected significantly by the presence of resistive payload in the generator output due to it has the greatest RMSE and RMS values regarding to the other disturbances.

Table 2. PID controller performance indexes in presence of external disturbances

Test	RMSE	RMS
Nominal operation	1	3,03
External disturbance	1,08	3,03
Random noise	1,09	3,05
Resistive payload	1,65	3,1

7. Conclusions

This paper presented the design and practical digital implementation of a PID controller applied to the speed control of a motor-generator system. The process was identified as a second order system with real poles. The PID controller was tuned employing the pole placement method and discretized through the backward difference discretization technique considering the implementation of the back calculation antiwind-up technique for the integral action. PID digital implementation was performed with the Matlab stateflow toolbox with a data acquisition card, showing a good performance for different operating points of the system. The PID controller performance was validated in presence of external disturbances, being the presence of resistive payload at the generator output that significantly affects the operation of the controller.

References

[1] Aström K J and Hägglund T 2009 *Control PID avanzado* (Madrid, España: Pearson Prentice Hall)

[2] Visioli A 2006 *Practical PID Control* (Springer London)

[3] Ribic A I and Mataurek M R 2016 Antiwindup PID controller design of TITO system based on ideal and inverted decoupling *2016 International Symposium on Industrial Electronics (INDEL)* pp 1–8

[4] Shin H B and Park J G 2012 Anti-Windup PID Controller With Integral State Predictor for Variable-Speed Motor Drives *IEEE Trans. Ind. Electron.* **59** 1509–16

[5] Wang Q, Wang C, An K, Shen G, Zhou X and Li Z 2014 Anti-windup PID position control of cylinders for a aircraft skin inspection robot with double-layer structure *Proceedings of the 33rd Chinese Control Conference* pp 8365–9

[6] Astrom K 1995 PID controllers: theory, design and tuning *Instrum. Soc. Am.* 343-

IOP Conf. Series: Materials Science and Engineering **437** (2018) 012014 doi:10.1088/1757-899X/437/1/012014

Productive chain management around disruptive technologies. Engaging sustainability and materials research

H V Martínez[1]

[1]Department of Mechanical Engineering, School of Engineering, Universidad Pontificia Bolivariana, Circular 1a 70 01, Medellín, Colombia

hader.martinez@upb.edu.co

Abstract. Mining is always a controversial issue and mainly when the discussion is addressed to developing countries. In Colombia, the mining panorama was and continues to be ruled by a formal, modern, and large-scale mining mainly focused in coal, the exploitation of nickel, and some companies that exploit gold in different places. Also, a large part of the mining activity is done at minor scales, defined in many cases by informality and the subsistence character. Bearing sustainability in mind, it is worth asking about the balance implies by the exploitation without added value of local and strategic materials that are irreversibly scarcer every time, including metallic and non-metallic minerals. Opportunities around natural and non-renewable resources must focus on enhancing the quality of life, the question addressed here is how to penetrate in the global markets of sophisticated products as well as how to be serious about the so-called "innovation", but with accent on sustainability and disruptive technologies.

1. Introduction

In order for a society to reach the capacity to be sustainable, certain economic, social, environmental, and energetic conditions must guarantee sufficiency of the available resources. This becomes a global and shared challenge in the context of different supply chains that consume and transform resources characterized by their scarcity. The development is sustainable if it involves a non-decreasing average life quality. Sustainability is a requirement of current generation to manage the resource base such that the average quality of life that societies ensure for themselves can potentially be shared by all future generations. In this context, sustainable development, the rational use of the resources, in particular non-renewable resources, and the evolution of technologies must be consistent with each other [1], especially in the current framework of history where the disruptive technologies (DTs) acquire an increasing relevance. DTs are technologies that broadly show the power of innovation to improve lives, transform industries and safeguard the planet. These technologies are generally new but include older technologies that are still controversial, relatively undeveloped in potential and do not have exclusively one application field. DTs include, but are no limited to artificial intelligence (AI), cloud computing, robotics, 3D printing (also known as additive manufacturing) and nanotechnology (NT), among others.

The general question addressed in this article is how to penetrate in the global markets of sophisticated products as well as how to be serious about the so-called "innovation", but with accent on sustainability. As a first consideration, the different types of materials in 3DP processes as well as different kind of nanomaterials are really forms of matter as metallic minerals and non-metallic minerals or industrial-minerals, which match two of the categories of the mining exploitation in Colombia. These categories are the first materials-type that could easily support the development of nano-intermediates, as long as 3DP technologies for processing a variety of polymers, metals, ceramics and composites [2]. The metallic and industrial minerals become of interest for different sectors and applications, besides having higher facilities for industrial appropriation in a changing and diverse market. This is why the appropriation of nanotechnology and 3DP makes sense in the context of a country with a traditional, artisan and large-scale mining like Colombia.

The true is that despite the higher level of foreign investment in recent years, the country continues to distinguish itself for its capacity for export raw materials. Conversely, the country mainly imports manufactured goods with more technology and added value. In other words, Colombian foreign trade continues to be tied to the dynamics of oil and mining activities. According to the Foundation for Higher Education and Development (Fedesarrollo), in the last 20 years the country has entered in a process of "re–primarization" of exports, sales of primary goods rose from 45% of the total in 1992 to nearly 70% to date, affecting industry growth and the creation of quality jobs. This process of deindustrialization could enter deceleration if, among other actions, agriculture and industry in general were further developed. Half of the Colombian GDP, estimated in 770 trillion in 2014, corresponds to the performance of four key sectors that besides financial establishments and commerce involve agriculture and industry. How to ensure that companies and these sectors – in the third world – move to the first world? In particular, Colombia is still far from achieving a real change in the composition of exports, where most products have added value. In a wider context, the first key is the generation, appropriation and dissemination of knowledge as a base for social and economic development, and also to close to exceed the gap for investment in R&D, which has reached less than 0,19% in proportion to the GDP [3]. In a study about economic growth in the six largest developing countries in Latin American, in which Colombia was included, Astorga [4] mentioned that despite significant advances in life expectancy and basic education, renewed efforts to strengthen the government's role in fostering secondary education and research activity remains a priority. Astorga also remarks that there is a need to secure macroeconomic stability to set the basis for high and sustained economic growth. This includes, among others, reinforcing and diversifying the export sector to minimize exposure to terms of trade fluctuations. The second key is diversification, which represents the element that correlates with scientific and technological competitiveness according to Cimini et al [5], who also mentioned that technological leading nations, beyond having the largest production of scientific papers and the largest number of citations, do not specialize in a few scientific domains [citar a F. Di trochio "las metiras de la ciencia" la ciencia no debe sesgarse: science must not be biased]. Rather, they diversify as much as possible their research system. On the other hand, less developed nations are competitive only in scientific domains where also many other nations are present.

2. Dynamics of commodities facing sustainability in Colombia

During the nineteenth century the comparative advantage of the products exported by Colombia resided in the abundance of some natural resources and unskilled labor, mainly in the agricultural sector and in mining. For example, hides, straw hats, platinum and tobacco, besides high concentration of exports in gold and coffee, this last one the only product that maintained the high export dynamism in this century. During the twentieth century, the share of exports other than coffee, crude oil, bananas and gold were remarkable just in a few periods. Currently, Colombia is recognized as a country with a high mining potential. In the world map of gross metallic ores extraction per country area, the country was ranked in more than 140 tons per km^2 in 2010 [6]. The mining sector has contributed historically close to 2,2% of the national GDP, and intends a participation 3,6% of the national GDP according to the Ministry of Mining and Energy. In 2012, the non- traditional exports of the mining sector were of USD 12,193 million, with an annual average of USD 13,765 million between 2015 and 2016, which show a record of historical growth (figure 1). Complementarily, the Direct Foreign Investment (DFI) accumulated in mining, which between 2005 and 2014 was of USD 33,770 million, is projected to USD 51,776 million between 2015 and 2025. Excluding the mining-energy sector, for industrial and construction minerals as well as for metallic minerals, starting in of 52% in 2015, the projections of DFI for 2020 and 2025 will be respectively 62% and 79%. In this sense, the mining sector has invigorated the national economy, generating large incomes from exportations, attracting DFI and it is the great funder of social programs. In parallel, the royalties from the mining sector have supported largely the regional and national infrastructure.

2.1. Non-traditional and high technology exports

Nevertheless, it must be highlighted that the Colombian mining panorama was and continues to be dominated by a formal, modern, and large-scale mining, but focused in the large coal projects located in the Caribbean coast and in the exploitation of nickel in Córdoba region. In contrast, a large part of the mining activity is done at minor scales, defined in many cases by informality and the subsistence character. Colombian mining employs around 350,000 people. 315,000 are dedicated to small and medium mining, with 76% of unqualified manpower, generating 59% of indirect employments and 41% of direct employments. On the contrary, large mining generates only about 35,000 employments, of which 56% are direct and 44% are indirect. Towards this unbalance there is a big question in terms of profitability and competitiveness, but mainly in terms of sustainability, much more when the inclusion of the named "mining machine" in the view of the Colombian State was broadly qualified as one of the main resources for the development of the country and its regions.

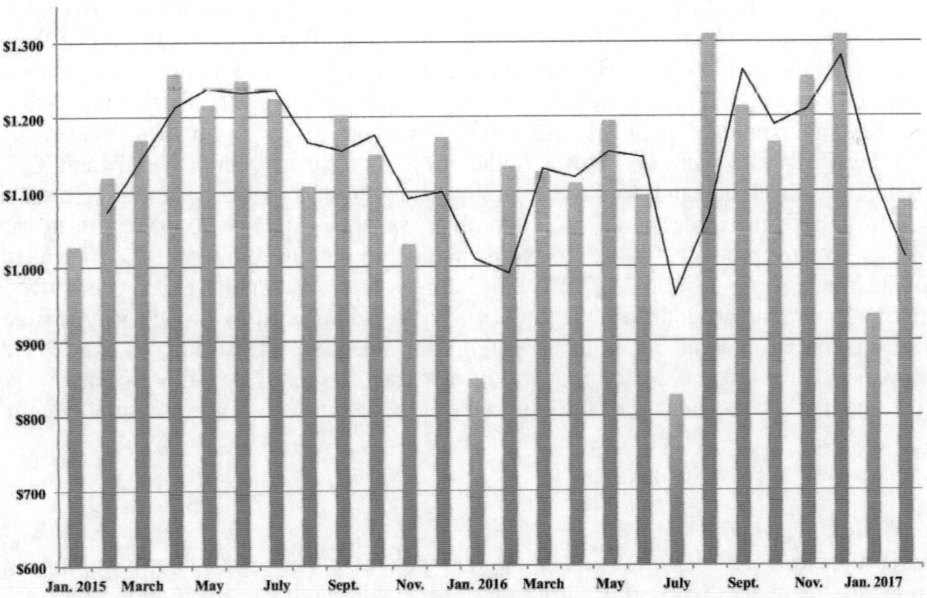

Figure 1. Not traditional exports of Colombia between 2015 and February-2017 (million USD) according to Mining analysis Bulletin from Colombian Ministry of Mines and Energy

As examples, in case of gold (Au), the region of Antioquia is the largest bulk gold producer in the country. This industry, together with the exploitation of other mineral resources such as coal, limestone, and construction materials, represent a significant source of royalties for this region since pre-Colombian times. Nevertheless, the gold activity settled in the culture and in the economy of many places in Antioquia contrasts with the technological gold products and productive chains established in developed countries, such as colloidal gold nanoparticles with unique optoelectronic properties useful in organic photovoltaics [7], sensory probes [8], therapeutic anticancer agents [9], drug delivery in biological [10] and medical applications [11] and catalysis [12], among others. As a second example, different metals such as cobalt, antimony, beryllium, fluorite, gallium, germanium, indium, magnesium, niobium, rare earth elements, tantalum, tungsten, including platinum (Pt) and the platinum group elements (PGEs), are also considered worldwide as critical main sources. In particular, because its physical and chemical properties, Pt is everyday more a “critical and environmental metal” than a precious metal, estimating that 20% of the products made in the world have Pt, mainly in the automobile industry, as the main component in screens for catalytic converters since the 80s decade, and used to transform toxic pollutants, such as carbon monoxide and nitrous oxide, into carbon dioxide, nitrogen, and water vapour. The use of Pt is therefore closely related to the demand for automobiles, and currently is very valuable, even more than Au, and the most important among the group of the nine most rare metals (that include Ru, Pd, Rh, Os, Ir, Pt; plus Au, Re, and Te) [13]. In Colombia, the region of Chocó currently produces the mainly bulk platinum in the country. Nonetheless, the Pt mining activity is informal in many cases [14]. Again, this situation is opposed to the technological products and productive chains established in developed countries around Pt and the PGEs, which are vital to fuel cells [15], as well as in the glass, petroleum, electric, and electronic industries, and the manufacture of jewelry. Besides, PGEs are exploited in medicine for cancer treatment and the preparation of dental fillings [16].

Actually, the economic growth in the last decade was possible in an exceptional manner in Colombia due to the increase of prices of certain mineral resources - mainly mining-energy resources-, which are highly sensible to the variation of the dollar price, the demand from developed countries, and clearly, to their general availability in the world [6]. In other words, Latin America is affected positively with a high price of the dollar, but only because it is a region that is highly dependent on the exportation of main sources, i.e. raw materials, oil and metals. But, with the arrival of the circular economy [17] and new supplies to world markets, the popularity of these main sources starts to finish and the prices of iron, copper, and especially oil, can dropped to low levels. According to the OECD, the end of the long period of popularity of the international prices of main sources would have concerns of high impact. Therefore, towards the need of a sustainable development for mining together with the global demand for more sophisticated products, the economic growth without productivity will have no sense. Even so, Colombia has not been able to penetrate in the world markets of sophisticated products (figure 2), and its natural non-renewable resources, such as minerals, are becoming increasingly depleted against a backdrop of extraction for export and low national consumption rates and intense industrialisation [18]. The previous reasons explain why it is necessary to be serious about innovation but with accent on sustainability. According to the World Trade Organization, between 1980 and 2010 the export of manufactures in the world grew an average of 8% yearly, more than the growth indexes of the exportations of main products (agricultural 5% and mining 6%).

2.2. Does mining sustainability exist?
By definition, the extraction of non-renewable natural resources denies the possibility of sustainable mining. In reality, there are diverse types of mining, some with high social and environmental impact and others with low impact. In contrast to the environmental impact of the extraction process, there is what is being done with the economic results that come from the mining activity, and how these are related or not to sustainable human development. It happens that countries with similar mining wealth generate completely different development processes, that depend on how the mining income is used, building, or not, the possibility to leverage the sustainable development with resources from extraction activities. In like manner, countries that export assets that are relatively sophisticated for their own development level grow faster that those which simply export more of the same traditional products, such as most of South American countries.

IOP Conf. Series: Materials Science and Engineering **437** (2018) 012014 doi:10.1088/1757-899X/437/1/012014

In Colombia there is attention to the mining royalties in social terms, related to covering the basic needs of the population in the health, education, drinking water, and basic sewerage sectors, among others. Nevertheless, the question remains what to do with informal mining, low scale mining, and even worse, with illegal mining that obviously does not pay any royalties? The effect of mining depends on how the activity is done and how it is framed in a long-term economic proposal and social policies. Consequently, besides the development of complex and sustainable productive capacities, the promotion policies for mining in the country must enable to add value to the mineral resources, increasing the operational efficiency and contributing to the mitigation of the environmental impact that would lead to a reduction in social conflict [20]. Also, it is imperious that Colombian exporters of non-renewable minerals and energy resources (coal and oil) integrate themselves to the world race for productive and sustainable innovation, as well as establishing highly efficient value-added chains. Otherwise, the current state will continue increasing the damage and scarcity of the available resources, without other guarantee different to the compensation through royalties and therefore, against the concept of sustainability.

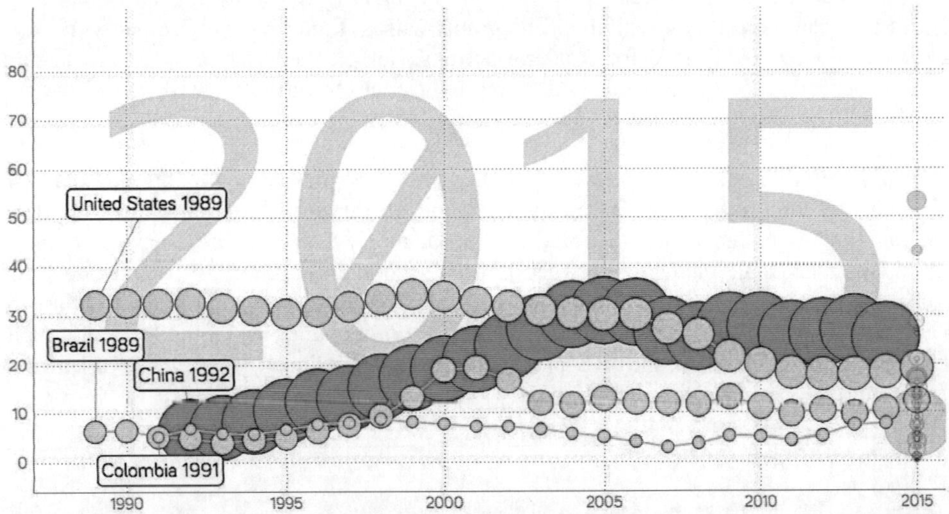

Figure 2. Comparative of High Technology Exportations (HTE) for the beginning of the XXI century. (% of exportations of manufactured products). Data simulated in Gapminder^{TM}. The size of the circles corresponds to the approximate size of the population according to the country. Colombia appears in the smaller circles and is compared to the U.S.A., China, and Brazil. HTEs refers to products with high intensity of innovation and development [19].

To have mineral resources and to use them is not by itself a negative activity. On the contrary, it can be a source of wealth and support for the sustainable well being of humans. Countries such as Sweden and Finland have done it; they have based their economic and social development partially on the finding and measured exploitation of mining resources. Other countries instead have exploited their mining resources to increase their consumption and very little to invest in the formation of human talent, without giving priority on the long term economic and social planning [21]. In Latin America, Chile has a fiscal rule that enables it to save lots of resources during the mining booms in order to spend them in bad times. This is why the mining countries, with high o low mineral sources require permanent and more R&D initiatives, in order to safely board a path towards development, adding the possibility to benefit "in a thoughtful and sustainable manner" from certain natural resources broad interest for diverse technological drivers in the world. One approach could consist in developing

IOP Conf. Series: Materials Science and Engineering **437** (2018) 012014 doi:10.1088/1757-899X/437/1/012014

methods that provide maximum use of the elements composing the ore and/or minimize the amount of residues.

3. Alternatives with added value

The worldwide productive structure is being oriented towards a society and an economy based on knowledge (that in other terms is the concept of modern societies), then the complementary problem for Colombia is to guarantee that the international competitiveness of the mining and industrial sectors will be based, not only on the exploitation, exportation of main sources and in the geological knowledge of the country, but also on the sustainable exportation with added value. The capacity of sustainably optimizing the exploitation and benefit of local resources can be increased, for example, through disruptive sciences in order to generate complex productive supply chains, minimizing the consumption of minerals, and offering innovative and mainly value-added products that progressively transferring the qualification of exporter of commodities and main sources, to the status of producer of intermediate assets, and even final assets, for diverse applications of higher economic recovery. Second, appropriating circular economies as links in productive chains for the reuse of resources, materials and products, will increase not only the efficiency, but also the sustainability in the framework of the scarce resources available. The globalization, Free Trade Agreements (FTA) and the open regionalism increase the need for strategic actions, but planned according to the sustainability criteria. A key aspect is the question of which applied–research initiatives will consolidate a model more centered on technology while combining disruptive sciences.

As far as disruptive sciences are concerned at least two examples can be mentioned here (See also figure 3): First, the NT global market has evolved and currently presents a greater opportunity to participate in the so–called nano–intermediate products, which are products that incorporate nanomaterials or nano–scale characteristics [53]. As a second example, 3DP will strongly influence manufacturing in aerospace, medical components, tooling and transport [22]. In particular, transport sector is and will be one of the biggest consumer for 3DP components, given the level of development required by different types of automotive–part suppliers and particularly for the companies that are part of the automotive supply chain in Colombia.

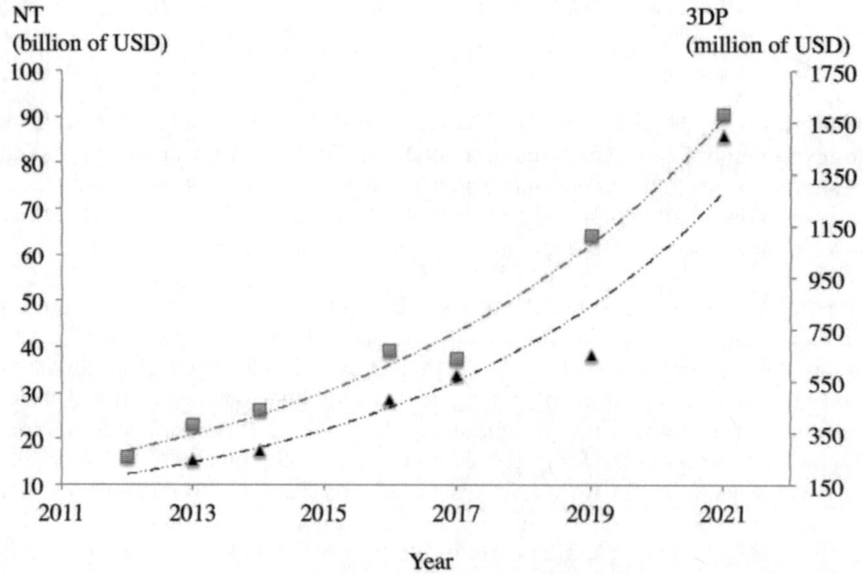

Figure 3. Global market forecast for NT (□) and advanced materials for 3DP, technologies and global markets (Δ). Data source: [23]–[25].

3.1. NT value chain

It is easier to sell something that is useful for different applications, with greater ease of appropriation in a changing and diverse market [54]. In relation to the sales of products that incorporate nanotechnology, it is convenient to distinguish three levels in the nanotechnology value chain (figure 4). The introduction of tools and instruments adapted to the work at the nanoscale level is an indispensable requirement to develop any application (Level 0). Therefore, the manufacturers of instruments for processing and characterization of nanomaterials, are the first segment of nanotechnology to reach the market. According to forecasts from the OECD, the growth in nanotools is expected to be around 30% annually. In 2005, nanotools market represented more than USD 266 million in the United States. The forecast is near to USD 10 billion in 2017. The next level is represented by the advancement of knowledge in nanomaterials (Level I), which industry develops to the large–scale supply of future applications [55]. The global market scale for nanomaterials is expected to double the numbers; with USD 15,9 billion in 2012, to 37,3 billion in 2017 [25] (figure 3). Level III is the entrance in the market of nanotechnology–derived applications, represented in products and processes, with the main target sectors being energy, health and electronics [13].

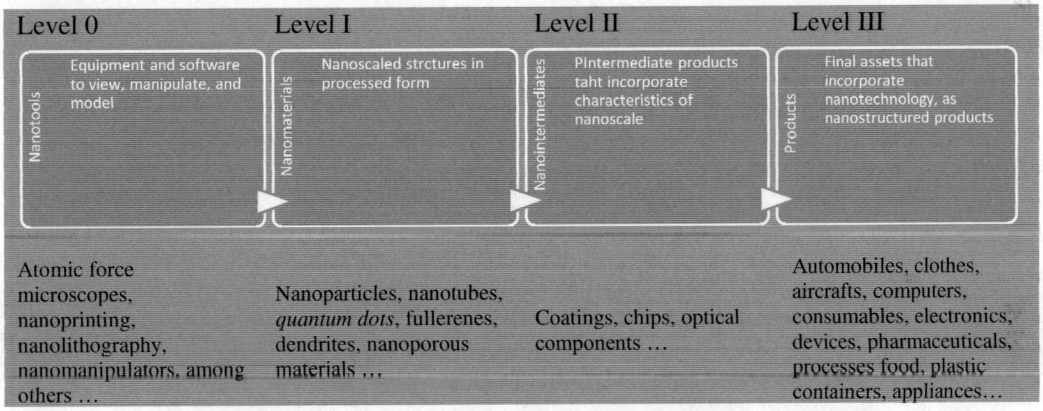

Figure 4. Nanotechnology value chain. Adapted from [26].

Indeed after the notable development of nanomaterials in the last 10–15 years, the challenge is currently to incorporate them into semi–finished products. Currently, the global nanotechnology market had evolved and presents a greater opportunity for the so–called nano–intermediate products (Level II). As a reference data, the relationship between applications for patents and publications increased from 0.23 in 1999 to 1.2 in 2008, thus indicating that besides discovery or basic research in nanotechnology, corporations have more interest toward commercial applications and benefits. In this context, not only in Colombia, but also in any emerging economy, the premises of applied research (pending and ongoing) directly related to nanotechnology and especially to nano–intermediate products should be the basis of projects, programs and activities that contribute to promoting national development. The appropriation of nanotechnology in combination with traditional, artisan, and large-scale mining has sense because some of the different types of nanomaterials are really forms of matter, such as oxides, nitrates, carbides, composites and metals. At the same time, they can be used in new developments and Colombian technological drivers, including structural and biological applications, photonics, medicine, and energy, among other topics that are having more commercial impact each time (table 1).

3.2. 3DP and transport

3DP allows objects to be fabricated layer by layer in a continuous or incremental manner, enabling three-dimensional objects to be 'printed' on demand [2], with significant advantages in design freedoms, mass customization, co-creation and innovative business models [41]. The global market for 3-D printing materials reached $475.4 million in 2015 and is expected to reach $576.6 million in 2016 and over $1.5 billion in 2021 (figure 3) [23]. The key concept suggested here is that when a part or spare is printed, mainly by using cyclical materials flows [42], the object will last longer. This is possible because 3DP shifts towards more localized production besides re-shift production to consumer countries as the share of labour costs in the total production costs decreases [22]. Also, despite its early maturation phase, it should become aware that 3DP represents a technology which can greatly lower the input and output intensities of industrial manufacturing [43] (figure 5).

Table 1. Level I of the value chain of nanotechnology. Examples of application of different types of nanomaterials obtained from different mineral sources

Nanoparticles	Application		Reference
Mn-Mg Ferrites	Space	Magnetic nanoparticles for spatial applications	[27]
SiO_2	Optics	Base covers with anti-reflective composites	[28], [29]
Gd, Pd	Optics	Exchangeable mirrors (Intelligent glass)	[30]
Ag	Optics	Films of different color	[31]
Ag	Biology and printed electronics	Conduction films obtained at low temperature	[32]
Ag	Biosensors	Quality Control in animal products, food and biological materials	[33]
Au	Medicine	Superficially functionalized nanoparticles with DNA for molecular diagnose	[34]
Carbon nanotubes	Catalysis	Decoration with Au nanoparticles Au, for catalytic applications	[35]
CuO	Catalysis	Diverse applications	[36]
$(1\text{-}x)\alpha\text{-}Fe_2O_3\text{-}SnO_2\ _{(x=0.0-1.0)}$	Sensors	Detection of gases	[37]
$xZrO_2\text{-}(1\text{-}x)\alpha\text{-}Fe_2O_3\ _{(x=0.1,\ 0.5)}$	Sensors	Detection of CO	[38]
Fe_3O_4	Biology	Ferro-flows as contrast mediums in biological tissues	[39]
ITO (90% wt In_2O_3 - 10% wt de SnO_2)	Electronics	Smartphones, GPS, LCD, OLED, solar panels	[40]

In particular, automotive sector is recognized worldwide as the spearhead of economic and social development because multiple spill overs onto a wide range of industrial fields, so consumer products and automotive manufacturing represent the largest potentially applicable markets for 3DP as they are predicted to have a combined share of 86% in the total manufacturing market [22]. In this context, it is worth noting that the Colombian automotive industry faces the challenge of appropriating 3DP processes for metallic materials, polymers and ceramics to develop in the country different components, and enhancing the products that are already produced nationally trough traditional manufacture process, including suspensions, steering, exhaust, transmission and cooling systems, friction materials, electric parts such as batteries and wiring, wheels, chassis racks, metal parts and various accessories, among others. Despite the complexity and non-linearity of the Colombian business cycle [44], according to DANE (the National Administrative Department of Statistics of Colombia), the Colombian automotive industry had contributed 4% of the total industrial production in the country, of which 1,1% corresponds to motorcycle assembly. The transport of goods and services is not only vital for the productive apparatus of Colombia; it is a social and economic dynamic, generating 10% of the jobs in the country and contributing COP 33.4 trillion to its GDP.

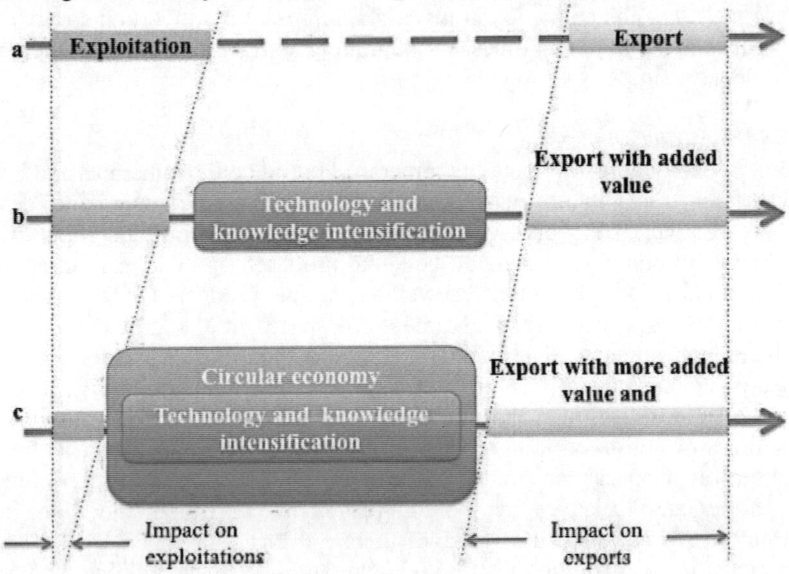

Figure 5. Productive local chain perspective around disruptive technologies as 3DP and NT in Colombia

The two alternatives exposed until this point are represented in figure 5, which include the traditional dynamics for exploitation and exportation of some commodities in Colombia in a first scenario (figure 5a). In a second scenario, technology intensification is done in NT and nano–intermediate products (figure 5b). In a third place, using 3DP of recycled materials and products a circular economy [17] is (complementary) constructed from production-consumption systems that maximizes the service produced from the linear and previous throughput flow (figure 5c). Figure 5 intends to indicate that in order to generate complex productive capacities that add value to non-renewable resources, and progressively migrate the status of a country from exporter of commodities and main sources, to that of producer of intermediate assets, and even final assets, it is important to invest in the design of technological supply chains (but also entrepreneurship, business models and high quality education), with the aim of formulating a research agenda to enable Colombia to reach its full potential for the future.

3.3. Local sustainability perspective on 3DP and NT for some mining resources in Antioquia

As a single and final example, between the 32 departments of Colombia the Department of Antioquia is most populated province and the country's largest economy after Bogotá. The investment in R&D is about 0,28% of its GDP, that even though it is slightly higher than the national investment, it is still low compared to international standards [45]. According to the Chamber of Commerce in Medellin, between 2001 and 2010 the exportations from Antioquia went from USD 2.006 million to USD 4.714 million (with a yearly growth rate made up by 10%). Nevertheless, the per capita GDP of Antioquia (without Medellín city and the Aburrá Valley) decrease to USD 3.685 per year, compared with the one of Medellín; which is about USD 6.105 yearly. This is caused by the productive structure in the most delayed parts of the Department, where besides the commerce, tourism, and the basic agriculture, there is mining exploitation. Technically and technologically, the mining activity in Antioquia is developed in a high percentage under inadequate conditions, resulting in the loss of deposits and a minimum management regarding environmental issues. For the department it is fundamental to reach a higher level of quality in the mining activity but with added value [46]. This makes sense if it is also considered that being Antioquia the larges gold producer of the country, the presence of other resources, as coal, limestone, and construction materials, represent an important source of the income for the department since pre-Colombian times, so the mining activity has settled in the culture and the economy of all the department.

3.3.1. Industrial and construction minerals

In Antioquia there is exploitation of metallic minerals, non-metallic minerals, and mining-energy resources. These different minerals are materials-type that could easily support the development of nano-intermediates and novel 3DP process in Antioquia. Notwithstanding, the chain of non-metallic minerals concentrates its production in cement, concrete products, and ceramic covers. Its exporting vocation is focused on glass and its products, as well as ceramic products for home use. Even though an important part of the sales are abroad, there is still space for the growth of exportations for industrial and construction minerals. The mineral rocks obtained in the quarries (marbles, granites, limestone, and slates, among others) can be useful for the development of coatings, made up by nanoparticles resistant to corrosion and the microbial flora. Together with the current popularity of construction, as sources of construction materials, in the Aburrá Valley the quarries in Bello on the Medellín-Bogotá highway, Copacabana, and Caldas are very well known. In the east of the department there are quarries in the San Nicolás Valley; including El Retiro, La Unión, and Rionegro. Regarding the quarries exploited for ornamental use (marble) there are locations in La Danta, and Río Claro. Marble is also extracted in Amalfi, Puerto Nare, Segovia, and Jordán. In Valdivia and Yarumal there are blocks of slate with grey to green tones know as Valdivia stones, of broad architectonical use in the department [47].

The greatest calcareous deposits are in Puerto Nare; which also include Puerto Berrío, Puerto Triunfo, San Luis, and Sonsón. The production of limestone, marble, and clay in 2010 was of 1097, 724.800, and y 720.000 tons respectively, with the district of Puerto Nare having an average production, projected to 2019 of 3.400.000 tons of limestone, according to the General Comptrollership of Antioquia [46]. This inventory accounts for the mineral deposits, evidencing the reserves of calcareous rocks (mainly marble), whose exploitation supplies, in a large measure, the cement industry in Antioquia, one of the most important in the continent. Such rocks are also exploited as ornamental rocks and as a source for agricultural and industrial salts. It is feasible to work in the direction of expanding the benefit and use, through NT, of minerals form Antioquia and Colombia, as supplies for the production of agrochemicals of interest in the fertilization of fields, as well as to improve the productivity and economic efficiency of the agricultural and farming activity in Colombia.

There is also an important number of commercial nanoparticles, derived from the same sources of industrial minerals that can be used to make a variety of nanocomposites. In this context, organic clays or nanostructured montmorillonita, or commonly denominated nanoclays, are the most researched nanoparticles for different polymer nanocomposites [48] in a broad range of promising 3DP applications [49]. Other uses of nanoclays are in the development of adsorbents, catalysts, and diverse biomaterials [50], [51] through 3DP [52]. Antioquia has been a known producer of plastic clays, aluminates, and diverse kaolins since the late 20th century. Natural clay (bentonite) in Antioquia, contains besides montmorillonita, glass, kaolin, quartz, zeolites, and carbonates. The main industries that demand clays and kaolins are the ceramic, paint, refractory, and cement industries. In Rionegro (Llanogrande) and in El Carmen de Viboral there is extraction of homogeneous mixtures of plastic and aluminous clays, both for ceramic and refractory use [46]. In the municipality of La Unión, besides clay for ceramic, there is also exploitation of kaolins that are submitted to different benefit and treatment processes, to generate products that are mainly used in paints and pigments, sanitary and electrical porcelain. Unprocessed kaolin is used in the cement industry. Besides these resources, there are red clays and slits of broad use in the construction industry as a main source for the production of bricks and roof tiles. In the municipalities of Valparaiso, Angelópolis, Amagá, Abejorral, and Santa Fe de Antioquia there is exploitation of compacted clays and other types of clays that are useful in the production of ceramic tiles, and in a lower proportion, for the production of bricks and roof tiles. Other clays with variable contents of iron, calcium, sodium, potassium, alumina, and silica, are of interest for the production of grey cement and are generally exploited close to the cement factories [47]. All these materials could be also useful for the additive manufacture of ceramic components [53], [54].

A third group of applications of nanomaterials and development of intermediate products is found in the talc mineral. This one can be used as a nano-additive for lubricant oils to reduce friction [55], or as nanoparticles that reinforce thermoplastic matrixes [56], among other uses. In this context, the municipality of Yarumal in Antioquia has traditionally been a producer of this mineral. Generally, the talc extracted is submitted to processes of trituration, milling, and classification and is used according to the level of purity in paints, pigments, plastics, soaps, rubbers, clays, papers, and ceramics, but not precisely in local applications related to nanotechnology.

Assisted by the current popularity of the civil industry, one of the construction materials that are mostly used for housing, besides limestone, dolomite, and other types of materials is plaster. This material is being studied by nanotechnology for the performance of innovations in the construction industry, especially because of its potential uses in systems for the control of pollution, self-cleaning mediums, and photo sterilization, through the addition of nanoparticles of titanium dioxide (TiO_2) [57]. In Antioquia, there is exploitation of a plaster deposit, formed by volcanic activity and located in the municipality of Anzá, throughout the Niverengo creek. The exploited material, depending on its purity, is used by the cement and construction industries, in this last case for the fabrication of stuccos and ornamental plasters [47] and could be used for cultural heritage applications [58] through 3DP. In general, the use of nanoparticles in cement materials results in a broad range of possible innovations, defined by their functionality and an improved mechanical performance, even with the potential energetic expense associated to new manufacturing techniques. The effect of chemical inertia of these type of materials, added to their growing consumption by the cement industry, is showing better results in the hydration times, concentration phenomena, and pre structure; favouring the levels of disinfection and tenacity [59]. These properties have been found to be modifiable and adaptable according to the need as a function of the size of the added nanoparticles (calcium and magnesium carbonates, clays, silica, unprocessed kaolin, among others). Complementarily, the analysis of the life cycle of different products show that the photo-catalytic reactivity of TiO_2, can compensate possible environmental damages [57]. With these conditions, the nanoparticles with base of limestone and dolomite, and other types of minerals, can also revolutionize the construction industry in Antioquia, through the creation of new cement based composite materials (CCM); adapted according to their structure and properties for different structural or functional requirements [60]. CCMs, reinforced with carbon nanotubes tell a recent story, in which the fabrication methods, the structural, mechanical, electrical, and piece-sensitive properties, as well as thermal and shock or bullet proof properties, continue to be the object of improvements, both in performance, and in cost, towards a broad range of

IOP Conf. Series: Materials Science and Engineering **437** (2018) 012014 doi:10.1088/1757-899X/437/1/012014

applications [61]. Here there are more opportunities for the development of nano-intermediate products from the mineral sources in Antioquia, as long as there are important reserves of limestone (calcium oxide, calcium carbonate) and dolomite (calcium and magnesium carbonate), mainly used for the cement industry. The cement plants are located in the east and northeast areas of the department, in the zones of Río Claro and Río Nare and in the southeast, in Abejorral and Santa Bárbara. There is also limestone in Sonsón, Amalfi, Cocorná, Remedios, Yarumal, San Carlos, Jordán, Puerto Berrío, and occurrences with possibility for exploitation in Río Verde and Togoridó, to the west of the department [47]. The limestone from Río Claro, besides cement, is considered a value-added product since it is destined for use in white paper for printing and writing, besides dull, varnished, cardboards and water based paints, varnishes, inks, stuccos, lacquers, plastics and rigid and flexible PVC products, hoses, cables, pharmaceutical and food products, dental paste, cosmetics, abrasives, water sealing products, soaps, and others.

Finally, regarding quartz sands (SiO_2), that are basically a sub-product of the benefit and treatment processes of kaolins, their industrial uses are broadly related to the ceramic and glass industries. In Antioquia there is exploitation of layers of hard sands in Angelópolis-Amagá and of pheldespate in La Ceja and Montebello, and in lower amounts, in Envigado, La Pintada, and Alejandría. The opportunities derived from nanotechnology regarding the use of SiO_2 include different fields, that go from water purification and advanced filtering means, medical and dental applications, development of composite materials, fabrication of glass products, the construction and abrasive tools sectors, among others [62], [57].

3.3.2. Metallic minerals

Copper, lead, iron, and zinc minerals have been extracted in the department of Antioquia in the mines of Zaragoza, Remedios, and Segovia. There are also indications of minor explorations and exploitations of copper and silver in La Ceja, and copper, molybdenum, lead, and zinc in Murindó, Urrao, Dabeiba, and Salgar. Other metals that have been exploited in the department in an organized or artisanal, isolated and interrupted manner are manganese in the La Loma creek in Santa Bárbara and in Dabeiba. In this case, the use of the manganese mineral is directly related to the brick and ceramic industry, but also, after being processes as manganese sulfate and potassium manganite, it has bee used to make fungicides in the chemical and pharmaceutical industries [47]. According to Colciencias, one of the reasons adduced by the experts for the low efficiency of agriculture (except for some exceptions such as sugarcane, flowers, or oil palm) and the Colombian farming activity, is the scarce use of adequate fertilizers and other agricultural supplies for the improvement of the crops and pastures. One of the main reasons is the price that most of these products, that are imported, tie their cost to the price of the energy needed for their extraction, processing, and transportation [63]. In such sense, it is feasible to work in towards the expansion of the benefit and use, through NT, of local and national minerals, as supplies for the production of agrochemicals of interest in the fertilization of the crops, and to improve the productivity and economic efficiency of the Colombian agricultural and farming activity. The use of mineral products of national origin, besides the manganese sulfates in Santa Bárbara and Dabeiba, added to phosphates, potassium chloride, sulfur, borax, dolomitic limestone, and other derivate from hydrocarbons such as urea and ammonia, have a strong justification for the development of the fertilizer and agricultural supplies industry in the country, due to the importance of their participation in the cost structure of the agricultural and farming sector, and to the low participation of the national industry in their production. Other applications for manganese sulfate and potassium manganite are found in the chemical and pharmaceutical industries, in water nano-filtering systems, in nanostructured catalyzers to eliminate bad smells and in the bleaching of textiles. Besides manganese, there have been chrome exploitations in the Santa Elena region, to the east of Medellín. Traditionally, this mineral has been used as foundry sand and in the production of colored vitreous chips. Chemical derivations of this mineral have also been used in leather treatment processes, as pigments for inks and paints, besides textile processing [47].

4. Conclusions

The mining opportunities must continue enhancing the life quality of the regions, aiming to position the country in other aspects, not only the exportation, but also the generation of knowledge and high levels of productive specialization. It is not an innovation, but is possible to imitate successful experiences from the strategies of some developed countries. So, it is not exaggerated to say that if Colombia takes good advantage of its mining, it can jump to sustainable development. The emphasis in the mining sector is a unique opportunity to give the country a push regarding infrastructure, and to end poverty to a large part of the 46% of Colombians that do not have enough income for a dignifying life. Particularly, in Antioquia Colombia, as well in many other regions in the country, there are clear opportunities for the rational benefit and with higher added value of different resources that can supply the large material demands implied by current and future technological developments. The essential ingredient in the exporting sophistication for mining is innovation in key of sustainability. Thus, the productive structure of diverse sub-regions in the department could be favourably impacted if they carry out projects of productive development in mining, together with the objective of developing nanomaterials as example, and looking for better levels of efficiency regarding the exploitation and the commercialization of resources with a higher added value through 3DP.

Acknowledgments

The author wishes to acknowledge to W.F. Flórez-Escobar for discussions of different paragraphs, the assistance of D. Nope-Beltran and the encouragement from D. A. Flórez-Londoño. The ideas and opinions expressed here are those of the author only and do not necessarily represent those of institution.

References

[1] D. Apelian, “Materials science and engineering’s pivotal role in sustainable development for the 21st century,” *MRS Bull.*, vol. 37, no. 4, pp. 318–323, Apr. 2012.

[2] V. Petrovic, J. Vicente Haro Gonzalez, O. Jordá Ferrando, J. Delgado Gordillo, J. Ramón Blasco Puchades, and L. Portolés Griñan, “Additive layered manufacturing: sectors of industrial application shown through case studies,” *Int. J. Prod. Res.*, vol. 49, no. 4, pp. 1061–1079, Feb. 2011.

[3] J. Lucio, D. Lucio-Arias, L.-A. Colorado, S.-C. Rivera, D.-A. Cruz, G. Usgame, G.-I. Perea, A. Guevara, É. Sánchez, F. Cifuentes, H. Mora, J. Cárdenas, J. Tique, M. Galvis, and V. Barón, “Indicadores de Ciencia y Tecnología 2012,” Bogotá, 2012.

[4] P. Astorga, “A century of economic growth in Latin America,” *J. Dev. Econ.*, vol. 92, no. 2, pp. 232–243, Jul. 2010.

[5] G. Cimini, A. Gabrielli, and F. Sylos Labini, “The Scientific Competitiveness of Nations,” *PLoS One*, vol. 9, no. 12, p. e113470, Dec. 2014.

[6] A. Schaffartzik, A. Mayer, N. Eisenmenger, and F. Krausmann, “Global patterns of metal extractivism, 1950–2010: Providing the bones for the industrial society’s skeleton,” *Ecol. Econ.*, vol. 122, pp. 101–110, Feb. 2016.

[7] R. C. Wadams, C. Yen, D. P. Butcher, H. Koerner, M. F. Durstock, L. Fabris, and C. E. Tabor, “Gold nanorod enhanced organic photovoltaics: The importance of morphology effects,” *Org. Electron.*, vol. 15, no. 7, pp. 1448–1457, Jul. 2014.

[8] V. K. K. Upadhyayula, “Functionalized gold nanoparticle supported sensory mechanisms applied in detection of chemical and biological threat agents: A review,” *Anal. Chim. Acta*, vol. 715, pp. 1–18, Feb. 2012.

[9] N. Singh and J. Abraham, “Gold Nanoparticles: Their Properties and Role as Therapeutic Anticancer Agents,” in *Nanoarchitectonics for Smart Delivery and Drug Targeting*, Elsevier, 2016, pp. 647–666.

[10] K. S. Siddiqi and A. Husen, “Recent advances in plant-mediated engineered gold nanoparticles

and their application in biological system," *J. Trace Elem. Med. Biol.*, vol. 40, pp. 10–23, Mar. 2017.

[11] M. Sengani, A. M. Grumezescu, and V. D. Rajeswari, "Recent trends and methodologies in gold nanoparticle synthesis – A prospective review on drug delivery aspect," *OpenNano*, vol. 2, pp. 37–46, 2017.

[12] G. J. Hutchings and J. K. Edwards, "Application of Gold Nanoparticles in Catalysis," 2012, pp. 249–293.

[13] Y. Kanazawa and M. Kamitani, "Rare earth minerals and resources in the world," *J. Alloys Compd.*, vol. 408–412, pp. 1339–1343, Feb. 2006.

[14] D. Tubb, "Muddy Decisions: Gold in the Chocó, Colombia," *Extr. Ind. Soc.*, vol. 2, no. 4, pp. 722–733, Dec. 2015.

[15] A. J. Appleby, "CHEMISTRY, ELECTROCHEMISTRY, AND ELECTROCHEMICAL APPLICATIONS | Platinum Group Elements," in *Encyclopedia of Electrochemical Power Sources*, Elsevier, 2009, pp. 853–875.

[16] Z. Robinson, "Sustainability of platinum production in South Africa and the dynamics of commodity pricing," *Resour. Policy*, vol. 51, pp. 107–114, Mar. 2017.

[17] J. Korhonen, A. Honkasalo, and J. Seppälä, "Circular Economy: The Concept and its Limitations," *Ecol. Econ.*, vol. 143, pp. 37–46, Jan. 2018.

[18] L. Gabriel Carmona, K. Whiting, A. Valero, and A. Valero, "Colombian mineral resources: An analysis from a Thermodynamic Second Law perspective," *Resour. Policy*, vol. 45, pp. 23–28, Sep. 2015.

[19] H. V. Martínez-Tejada and O. Muñoz, "Opportunities and Challenges for the Colombian Platinum: A Geopolitical Analysis of Platinum Mining in Colombia," *Analecta Política*, vol. 4, no. 5, pp. 387–410, 2013.

[20] S. Arango-Aramburo, P. Jaramillo, Y. Olaya, R. Smith, O. J. Restrepo, A. Saldarriaga-Isaza, J. Arias-Gaviria, J. F. Parra, E. R. Larsen, L. M. Gomez-Rios, and L. Y. Castellanos-Niño, "Simulating mining policies in developing countries: The case of Colombia," *Socioecon. Plann. Sci.*, Apr. 2017.

[21] P. Söderholm and N. Svahn, "Mining, regional development and benefit-sharing in developed countries," *Resour. Policy*, vol. 45, pp. 78–91, Sep. 2015.

[22] M. Gebler, A. J. M. Schoot Uiterkamp, and C. Visser, "A global sustainability perspective on 3D printing technologies," *Energy Policy*, vol. 74, pp. 158–167, Nov. 2014.

[23] A. McWilliams, "Advanced Materials for 3D Printing: Technologies and Global Markets," BCC Research, 2016.

[24] A. McWilliams, "Nanotechnology: A Realistic Market Assessment," BCC Research, 2014.

[25] A. McWilliams, "The Maturing Nanotechnology Market: Products and Applications," BCC Research, 2016.

[26] A. Vicari, "How to Combine Nanotech with Business Success," in *Nano MTY 2013. Nanotechnology Industrial Applications*, 2013.

[27] S. K. Sharma, R. Kumar, S. N. Dolia, V. V. Siva Kumar, and M. Singh, "Magnetic nanoparticles for space applications," *MRS Proc.*, vol. 851, p. NN10.4, Feb. 2011.

[28] Y. Hoshikawa, H. Yabe, A. Nomura, T. Yamaki, A. Shimojima, and T. Okubo, "Mesoporous Silica Nanoparticles with Remarkable Stability and Dispersibility for Antireflective Coatings," *Chem. Mater.*, vol. 22, no. 1, pp. 12–14, Jan. 2010.

[29] R. Serna, "Mesoporous Silica Nanoparticles Facilitate Antireflective Coating Applications," *MRS Bull.*, vol. 35, no. 2, p. 112, Jan. 2011.

[30] I. Aruna, B. R. Mehta, and L. K. Malhotra, "Application of Gd and Pd nanoparticles in a 'new generation' switchable mirror," *MRS Proc.*, vol. 888, pp. 888-V05-5, Feb. 2011.

[31] S. Hashimoto, T. O. Hirano, O. Okitsu, M. Ebisawa, T. Suzuki, and S. Maeda, "Ag Nanoparticle Films for Color Applications," *MRS Proc.*, vol. 1343, p. mrss11-1343-w05-03, Sep. 2011.

[32] K. Balantrapu and D. V. Goia, "Silver nanoparticles for printable electronics and biological applications," *J. Mater. Res.*, vol. 24, no. 9, pp. 2828–2836, Feb. 2011.

[33] D.-H. Tsai, S.-H. Guo, R. J. Phaneuf, and M. R. Zachariah, "Electrostatically Directed Assembly of Silver Nanoparticles for Application to Metal Enhanced Fluorescence Biosensing," *MRS Proc.*, vol. 951, pp. 951-E12-2, Feb. 2011.

[34] C. S. Thaxton, N. L. Rosi, and C. A. Mirkin, "Optically and Chemically Encoded Nanoparticle Materials for DNA and Protein Detection," *MRS Bull.*, vol. 30, no. 5, pp. 376–380, Jan. 2011.

[35] W. Lü, X. Ma, N. Lun, and S. Wen, "Decoration of Carbon Nanotubes with Gold Nanoparticles for Catalytic Applications," *MRS Proc.*, vol. 820, p. O3.10, Mar. 2011.

[36] M. Hosseinpour, S. J. Ahmadi, T. Mousavand, and M. Outokesh, "Production of granulated-copper oxide nanoparticles for catalytic application," *J. Mater. Res.*, vol. 25, no. 10, pp. 2025–2034, Jan. 2011.

[37] M. Sorescu, L. Diamandescu, and D. Tarabasanu-Mihaila, "Tin-Doped Hematite Nanoparticles for Gas-Sensing Applications," *MRS Proc.*, vol. 900, pp. 900-O05-3, Feb. 2011.

[38] M. Sorescu, L. Diamandescu, A. Tomescu, and S. Krupa, "Designing Mixed Oxides Magnetic Nanoparticles for Sensing Applications," *MRS Proc.*, vol. 1118, pp. 1118-K03-3, Mar. 2011.

[39] D. K. Kim, W. Voit, W. Zapka, B. Bjelke, M. Muhammed, and K. V. Rao, "Biomedical application of ferrofluids containing magnetite nanoparticles," *MRS Proc.*, vol. 676, p. Y8.32, Mar. 2011.

[40] J. E. Song and Y. S. Kang, "Synthesis of Indium Tin Oxide Nanoparticles and Application to Near IR-reflective Film," *MRS Proc.*, vol. 818, p. M5.14.1, Mar. 2011.

[41] T. Rayna, L. Striukova, and J. Darlington, "Co-creation and user innovation: The role of online 3D printing platforms," *J. Eng. Technol. Manag.*, vol. 37, pp. 90–102, Jul. 2015.

[42] M. Despeisse, M. Baumers, P. Brown, F. Charnley, S. J. Ford, A. Garmulewicz, S. Knowles, T. H. W. Minshall, L. Mortara, F. P. Reed-Tsochas, and J. Rowley, "Unlocking value for a circular economy through 3D printing: A research agenda," *Technol. Forecast. Soc. Change*, vol. 115, pp. 75–84, Feb. 2017.

[43] C. Feldmann and A. Pumpe, "A holistic decision framework for 3D printing investments in global supply chains," *Transp. Res. Procedia*, vol. 25, pp. 677–694, 2017.

[44] S. Aparicio, D. Urbano, and D. Gómez, "The role of innovative entrepreneurship within Colombian business cycle scenarios: A system dynamics approach," *Futures*, vol. 81, pp. 130–147, Aug. 2016.

[45] "Plan de Desarrollo de la Gobernación de Antioquia. Línea Estratégica: La Educación Como Motor de la Transformación de Antioquia," Medellín-Col., 2012.

[46] V. Villa Posada and G. Franco Sepúlveda, "Diagnóstico Minero y Económico del Departamento de Antioquia," *Boletín Ciencias la Tierra*, vol. 33, pp. 125–134, 2013.

[47] J. E. López-Rendón, "Recursos Naturales no Renovables y Minería," in *Geografía de Antioquia*, M. Hermelin, Ed. Medellín-Col., 2006, pp. 151–159.

[48] S. I. Marras, K. P. Kladi, I. Tsivintzelis, I. Zuburtikudis, and C. Panayiotou, "Biodegradable polymer nanocomposites: the role of nanoclays on the thermomechanical characteristics and the electrospun fibrous structure.," *Acta Biomater.*, vol. 4, no. 3, pp. 756–65, May 2008.

[49] A. C. de Leon, Q. Chen, N. B. Palaganas, J. O. Palaganas, J. Manapat, and R. C. Advincula, "High performance polymer nanocomposites for additive manufacturing applications," *React. Funct. Polym.*, vol. 103, pp. 141–155, Jun. 2016.

[50] P. Pasbakhsh, G. J. Churchman, and J. L. Keeling, "Characterisation of properties of various halloysites relevant to their use as nanotubes and microfibre fillers," *Appl. Clay Sci.*, vol. 74, pp. 47–57, Apr. 2013.

[51] C. H. Zhou and J. Keeling, "Fundamental and applied research on clay minerals: From climate and environment to nanotechnology," *Appl. Clay Sci.*, vol. 74, pp. 3–9, Apr. 2013.

[52] R. D. Farahani, D. Therriault, M. Dubé, S. Bodkhe, and M. Mahdavi, "Additive Manufacturing of Multifunctional Nanocomposites and Composites," in *Reference Module in Materials*

Science and Materials Engineering, Elsevier, 2017.

[53] A. Goulas and R. J. Friel, “Laser sintering of ceramic materials for aeronautical and astronautical applications,” in *Laser Additive Manufacturing*, Elsevier, 2017, pp. 373–398.

[54] B. Derby, “Additive Manufacture of Ceramics Components by Inkjet Printing,” *Engineering*, vol. 1, no. 1, pp. 113–123, Mar. 2015.

[55] P. Rudenko and A. Bandyopadhyay, “Talc as friction reducing additive to lubricating oil,” *Appl. Surf. Sci.*, vol. 276, pp. 383–389, Jul. 2013.

[56] L. Castillo, O. López, C. López, N. Zaritzky, M. A. García, S. Barbosa, and M. Villar, “Thermoplastic starch films reinforced with talc nanoparticles.,” *Carbohydr. Polym.*, vol. 95, no. 2, pp. 664–74, Jun. 2013.

[57] M. Vittoriadiamanti and M. P. Pedeferri, *Nanotechnology in Eco-Efficient Construction*. Elsevier, 2013.

[58] C. Balletti, M. Ballarin, and F. Guerra, “3D printing: State of the art and future perspectives,” *J. Cult. Herit.*, vol. 26, pp. 172–182, Jul. 2017.

[59] H. Sun, B. Hohl, Y. Cao, C. Handwerker, T. S. Rushing, T. K. Cummins, and J. Weiss, “Jet mill grinding of portland cement, limestone, and fly ash: Impact on particle size, hydration rate, and strength,” *Cem. Concr. Compos.*, vol. 44, pp. 41–49, Nov. 2013.

[60] A. R. Jayapalan, B. Y. Lee, and K. E. Kurtis, “Can nanotechnology be ‘green’? Comparing efficacy of nano and microparticles in cementitious materials,” *Cem. Concr. Compos.*, vol. 36, pp. 16–24, Feb. 2013.

[61] B. Han, X. Yu, and J. Ou, *Nanotechnology in Civil Infrastructure*. Berlin, Heidelberg: Springer Berlin Heidelberg, 2011.

[62] S. Jain, A. P. Jain, S. Jain, O. N. Gupta, and A. Vaidya, “Nanotechnology: An emerging area in the field of dentistry,” *J. Dent. Sci.*, Dec. 2013.

[63] H. V. Martínez-Tejada, S. Montoya, F. Jaramillo-Isaza, and M. Lainez-Álvarez, “Nanotecnología para Colombia. Una visión histórica, pasando por el contexto global, latinoamericano y las regiones,” *Rev. Nano Cienc. y Tecnol.*, vol. 2, no. 1, pp. 49–64, 2014.

IOP Conf. Series: Materials Science and Engineering **437** (2018) 012015 doi:10.1088/1757-899X/437/1/012015

Superficial modification by alkalization of cellulose Fibres obtained from Fique leaf

R E Guzmán[1], S Gómez[1], O Amelines[1] and G M Aparicio[2]

[1]Facultad de Ingeniería Mecánica, Universidad Pontificia Bolivariana, Campus Universitario Autopista Piedecuesta Km. 7, Bucaramanga, Colombia
[2]Ginsai Group, Universidad Autónoma de Occidente, Cali, Colombia

rolando.guzman@upb.edu.co

Abstract. The present study will focus on superficial modification by alkalization of cellulose fibres obtained from Fique leaf as reinforcement in polymer matrices to produce a natural fiber composite that can be suitable for industrial purposes. Fique fiber is a hard vegetable fiber derived from a Colombian plant (Agave furcarea). An appropriate treatment, namely alkalization, of the fiber was chosen and carried out, but will be customized for this specific study to be more environmental-friendly and economical. A higher fiber-to-solution ratio as well as a low concentration would decrease the price of the treated fibers. The changes introduced to the surface morphology by the abovementioned treatment is then examined using a scanning electron microscope (SEM), Thermogravimetric analysis (TGA), Differential scanning calorimetry (DSC) and FTIR analysis, and untreated and treated samples were.

1. Introduction

Due to the increase of the environmental consciousness around the world and with the purpose of diminish the damage to the environment, research of composite materials reinforced with natural fibers (NFC) has been boost, so that they can replace those of fossil origin, empowering the use of natural resources to accomplish this goal. Fique fiber is a hard vegetable fiber derived from a Venezuela and Colombian plant (Agave furcraea ssp.) whose production has expanded to other countries like Brazil, Ecuador, Cuba, Mexico, Costa Rica as well as other regions in Central and South America [1]–[3]. Since the fiber has high availability, low cost and is totally recyclable and biodegradable, it possesses technological, environmental and economic advantages. However, there are also certain limitations to its applications such as low mechanical properties and toughness; property variations; sensitivity to moisture, temperature and UV radiation; and fire resistance. Firstly, the fique fibers just like for other vegetable fibers, incorporation with an organic polymer matrix poses the greatest challenge due to the difference in surface nature. Polar OH groups are present in the chemical components, such as cellulose or hemicellulose, which makes these fibers highly hydrophilic. These hydroxyl groups and other oxygen containing groups are able to attract moisture through hydrogen bonding. Thus, they are prone to water absorption, leading to changes in fiber dimensions [4]. Since the polymer matrices are hydrophobic, good wetting of the fiber is hard resulting in poor adhesion at the fiber-matrix interface. The adhesion between fiber-matrix is the main factor that determines the mechanical properties of the composite material. Poor adhesion causes the load transfer from the matrix to the fiber to be ineffective, which leads to poor mechanical properties of the resultant composite [5], [6]. Application of chemical agents to the fiber and modification of the matrix could overcome this challenge [7]. Other disadvantages associated with the use of fibers are quality variations and low thermal stability [8]. Several investigations concerning to realize superficial modification to the hard natural fibers of reinforcement have been realized in sisal fibers [9]–[14], jute [15]–[19], bamboo [20]–[23], kenaf [24]–[27] and hemp [28], [29].

1.1 Natural fibres (NF)

The natural fibers are composed mainly of cellulose (Figure 1), which is a high resistance and rigidity polymer, composite and other compounds such as lignin, waxes and water-soluble compounds. Among its advantages it highlights its low price and weight and high resistance, similar to the synthetic glass fibers. Disadvantages, the hydrophobic polar nature of the natural fibers causes a bad adherence with a thermoplastic matrix and low microbial resistance.

Cellulose serves as the main component and provides the strength and stability in the plant. The glucan (polysaccharide of D-glucose monomers) consists of β-(1→4) linked monomer units. Cellulose is highly anisotropic with a theoretical strength of 10 GPa in the chain direction due to the molecular linearity. The pyranose rings are in the $^4C_1$ conformation, which means that the $-CH_2OH$ and $-OH$ groups, as well the glycosidic bonds, are equatorial with respect to the mean planes of rings [30]. Natural fibers are composed of a hierarchical structure [31]:

i) Vascular bundles of elementary fibers (fibrils)

ii) Cellulose microfibril bundles consisting of packed bundles of cellulose chains, in a network of hemicellulose and lignin.

The microfibrils are helically wound along the fiber axis and form hollow cells. The tensile strength of natural fibers are mainly owed to these microfibrils and its cellulose type, because each type of cellulose has its own cell geometry.

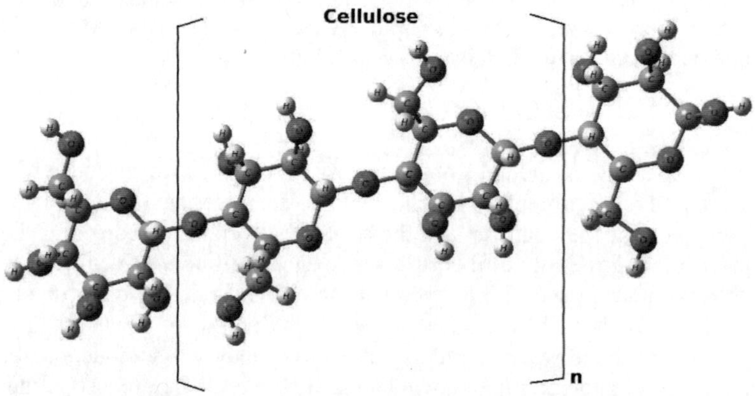

Figure 1. Chemical structure of cellulose chain

1.2 Treatment of natural fibers

There are several methods available to the enhance adhesion through chemical surface treatment of natural fibers involve silanization [32]–[39], Alkalinization[5]–[7], [14], [29], [31], [32], [37], [40]–[42], [42]–[60] , Esterification [6], [42], [61]–[64] benzoylation [10], [65]–[67], maleated coupling agents, isocyanate treatment, and grafting of synthetic polymers. The simplest way is by drying of the fiber before addition of the matrix. By doing so, the bonding of water molecules to the fiber surface is cancelled. More advanced treatments such as fiber surface modification, matrix modification, corona discharge, use of additives and coupling agents can enhance fiber-matrix adhesion. Surface modification by alkalization reduces the ability of cellulose to bond hydrogen, which also cancels the bonding of open hydroxyl groups with water molecules. In other words, it reduces the adsorption of water. Application of additives can be done on the fibers or the matrices. Coupling agents are compatibilizers between fibers and matrix. These agents remove weak boundary layers and create a cross-linked region in which there are covalent bonds between the matrix and fibers [68]. These treatments are capable of changing the chemical composition, physical properties, thermal stability, and surface structure.

1.2.1 Alkalization. Alkalinization, also known as mercerization or alkali treatment, is one of the most common treatments applied on natural fibers [23][25][64][69][6][41]. It is an irreversible process which involves the immersion of plant fibers in a relatively concentrated aqueous solution of a strong base for a set amount of time. The most commonly used base in this treatment is sodium hydroxide, NaOH. The treatment is mainly employed as a treatment of natural fibers that are to be used as reinforcements in a polymer matrix, thermoplastics and thermosets alike. The surface is cleaned by attacking non-cellulosic components in the fiber, which also increases the cellulose amount. Impurities and cementitious materials covering the external surface like lignin, wax, oils and residue from leaf not extracted previously are partly removed. Depolymerization of cellulose changes its fine structure and short length crystallites are produced. The surface roughness is increased through the disruption of hydrogen bonding in the network structure. More cellulose is exposed at the surface and there is an increase in fiber fibrillation. Thus, more reaction sites are produced, providing better mechanical interlocking with a matrix and, hence, better interfacial bonding. However, increased surface roughness could also promote the presence of voids in the resultant composites [42]. The treatment changes the fine structure, dimension, morphology and mechanical properties of the fiber. The mechanical behavior of natural fibers, more importantly the strength and stiffness, is significantly improved, which in turn also improves the mechanical properties of the resulting composite. Immersion of the natural fiber in aqueous hydroxide solution results in the ionization of the hydroxyl group to the alkoxide, which proceeds through the following reaction [4], [6], [7].

$$Fiber - OH + NaOH \rightarrow Fiber - O - Na + H_2O \quad (1)$$

The effect on mechanical properties by alkali treatment can be further enhanced by the application of tensile load [31]. Tensile load application during the treatment induces changes in microfibril orientation, resulting in an improvement in structural and mechanical properties. The cellulose microfibrils are the main contributors to the tensile strength of natural fibers. In addition, the spiral angle affects the mechanical properties due to the highly anisotropic nature of cellulose. The mechanical properties (have been found to) become higher when the orientation of the microfibrils are along the axis of tension. Treatment with alkali solution while under tension leads to the rearrangement of the microfibrils in the direction of tensile application. There is a decrease in the spiral angle (angle between fibrils and fiber axis) and increase in molecular orientation, which in turn would cause better load sharing and higher stress development in the fiber. A study on Ramie fibers showed that treatment under 0.098 N of tension with 15 wt% NaOH solution for 2 h increased their tensile strength compared to untreated fibers whereas slack alkali treated fibers had a decrease of 1.8 % (15). This procedure could help achieve greater mechanical properties, but with lower concentrations of NaOH.

2. Experimental results and discussion

2.1 Materials

The fique fibers were provided by Cohilados Ltda. The chemical agents used in this study were: Sodium hydroxide pellets (99%, Emsure) and glacial acetic acid (100%, Emsure). Distilled water was used for preparation of sodium hydroxide and acetic acid solutions. Fique fiber was cut to a length of about 5 cm and a total amount of 8 g was placed inside the glass container. The ratio of fiber weight to alkali solution was set to 1:10 for all treatments. The fiber was immersed in NaOH solution at room temperature for 30 min. After the set immersion time, the spent solution was decanted and the fibers were repeatedly washed with distilled water containing a small amount of acetic acid to remove traces of alkali solution.

IOP Conf. Series: Materials Science and Engineering **437** (2018) 012015 doi:10.1088/1757-899X/437/1/012015

To speed up neutralization, the sample was washed under running water and thereafter rinsed with distilled water. In order to remove residual moisture, the treated samples were dried in an oven at 60 °C until constant weight was reached (~ 6 hrs). The treated samples along with samples not subjected to alkalization to the fiber at different stages of the alkalization process are presented in Figure 2. Some samples were subjected to a pretreatment using fabric softener to ease the removal of surface particles.

Table 1. List of fique fiber samples

Samples of fiber	Treatment (softener)	Concentration wt% NaOH	Reaction time (h)
Sample A	Untreated	0	-
Sample B	Treated	0	24
Sample C	Untreated	2.5	1/2

2.2 Thermogravimetric (TGA) and Differential scanning calorimetry (DSC) Analysis

In order to know the thermal stability of the fique fibers, it proceeded to measure the behaviour of the mass of the fibers in function of the temperature, it's measured the fique natural fiber, the fique fiber that passed through concentrations of NaOH and the fique fibers passed through softener. All the measures were realized at a speed of 10°C/min. The samples (see table 1) were submitted to controlled heat from 25°C to 500°C until the process of degradation and combustion of the solid residues initiates, finally remaining the generated ashes.

As examples are shown in figure 2, the TGA and DSC analysis for the natural fique fibers (sample A) it's noted that as the fibers is heated, when it reaches ~150°C it has already lost approximately 7.045% of its weight, which can be attributed to the humidity in the material, once this humidity no longer existed, the fibers do maintain steady until approximately 280°C, temperature at which the process of decompose of the fique fibers starts, losing 71.34% between 280°C and 380°C, once 380°C is reached it initiates the degradation and combustion process, leaving 20% of the initial weight. DSC analysis was carried out to observe the behavior of the natural fique fibers (sample A), it presents a transition step type around -7.02°C which it is attributed to the vitreous transition of the fiber with an energetic requirement of 2J/g, then it is observed a transition of endothermic character that initiates at 10°C and finalizes around 130°C which can be attributed to dehydration of the fiber with a required energy of 247,3 J/g and ends with a transition of exothermic character around the 301.28°C with an energy liberation of the order of 15.93 J/g, that can be attributed to a possible combustion of the material for the high temperature reached. Mostly the thermal behavior of fique fibers presents three lose weight regions, which are show in Table 2.

The natural fibers fique (sample A) and treated fibers fique (Sample B) using softener has a maximum degradation about 360 °C and 355 °C respectively, whereas the fique fiber after 30 minutes of immersion in alkali solution (Sample C) has a maximum degradation al 337 °C. From figure 2.b), it can be seen that the treated fiber with fabric softener starts to lose weight earlier that samples B and Sample C. This may be attribute to the using fabric softener to ease the removal of surface cellulose particles. For the untreated fique fibers (without NaOH) two obvious peaks can be observed. The first peek at 240 °C corresponds to the decomposition of hemicellulose while the second peak at 355°C is due to the decomposition of cellulose.

IOP Conf. Series: Materials Science and Engineering **437** (2018) 012015 doi:10.1088/1757-899X/437/1/012015

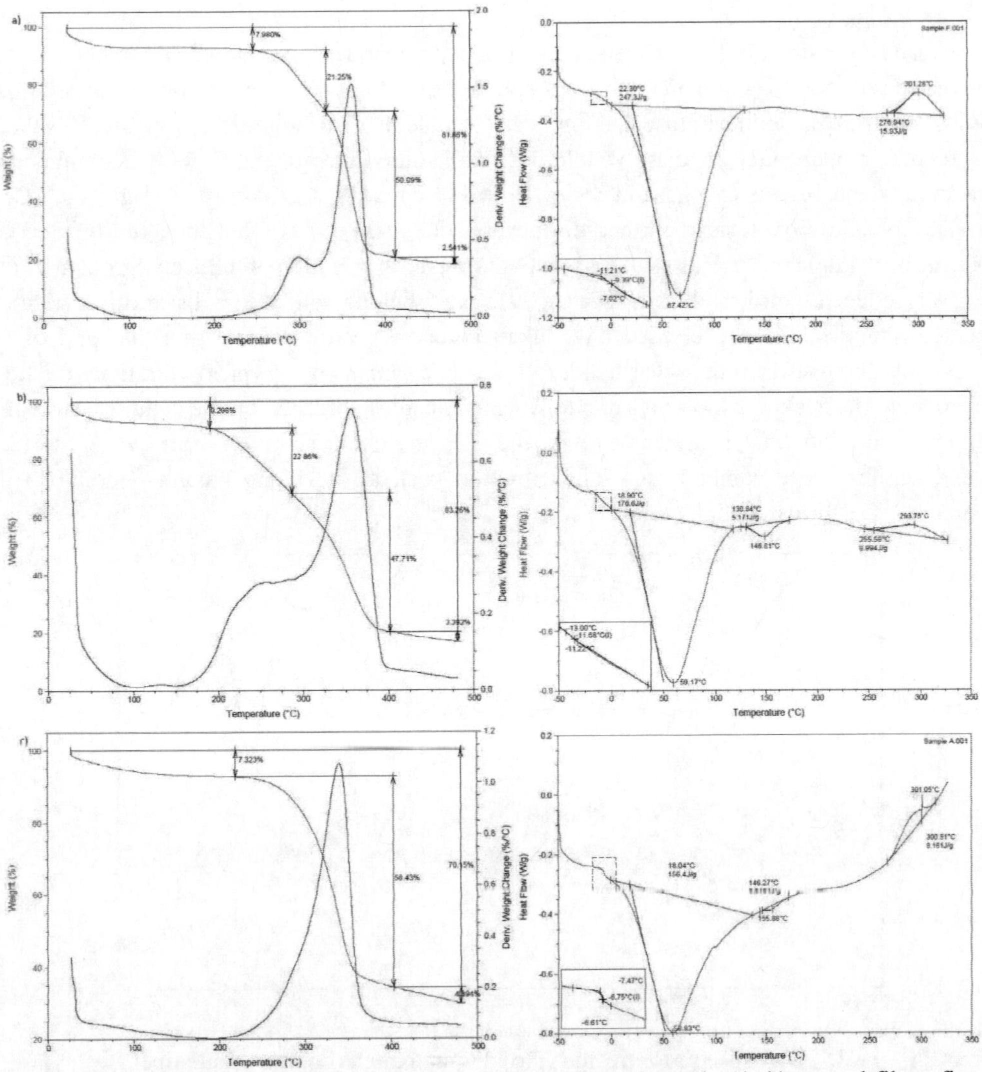

Figure 2. TGA/DTG and DSC curves a) natural fibers fique (sample A), b) treated fibers fique using softener (Sample B), c) fique fiber after 30 minutes of immersion in alkali solution (Sample C)

Table 2. Thermogravimetric results

Samples of fibre	T_1 (°C)	T_2 (°C)	T_3 (°C)	Weight loss (%)				Max. temp degradation
				Temp. T_1(°C)	Temp. T_2(°C)	Temp. T_3 (°C)	Mas weight loss(%)	
Sample A	240	332	380	7.980	29.26	81.32	81.86	361
Sample B	176	280	385	7.045	32.15	79.86	83.26	355
Sample C	236	255	365	7.323	-	65.753	70.15	337

2.3 FTIR Analysis

Fig. 3 shows band intensities in the infrared spectra of natural fique fibers untreated (sample A) and fique fiber treated with NaOH (sample C). It is well known that natural fiber mainly consists of cellulose, hemi-cellulose and lignin. Hemicellulose and lignin are soluble in alkali solutions; therefore, it is likely that a part of both components were dissolved during alkali solution treatment. Careful examination of the spectra revealed that several changes did occur. The vibration peak at 1731 cm−1, assigned to a C=O stretching vibration of carboxylic acid or ester, disappeared due to the removal of hemicellulose. Second, the vibration peak at 1240 cm^{-1}, which belongs to a C-O stretching vibration of the acetyl group in lignin component, was reduced. Third, a vibration peak at 2915 cm^{-1}, belonging to the C-H stretching vibration in cellulose and hemicellulose, decreased after alkali solution treatment indicating that part of the hemicellulose was removed. On the other hand, in both spectrums appear typical vibrations for fique fibers, for example the peak at 3330 cm-1 due to water molecules absorbed on the cellulose structure. The band located at 2850 cm-1 is associated with the C-H stretching vibration from CH and CH2 in cellulose and hemicellulose components. The vibration peak at 1026 cm−1 corresponding to the vibration of C–O–C (glycosidic linkage).

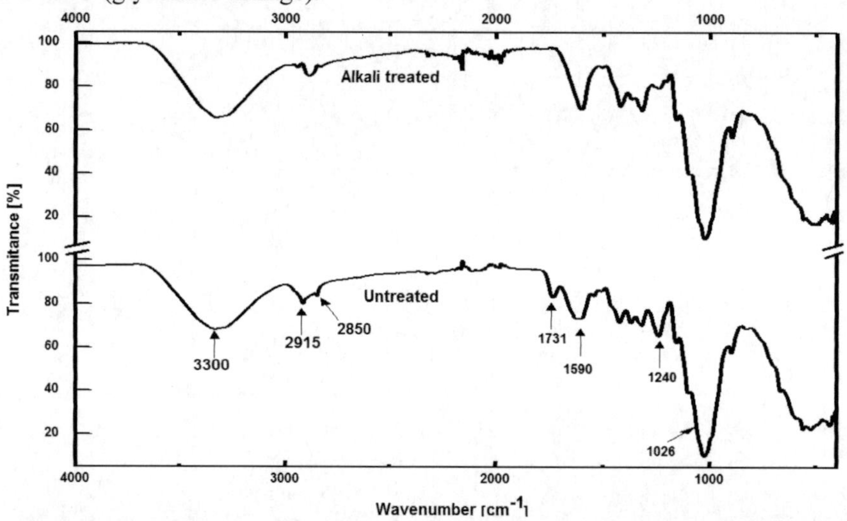

Figure 3. FTIR spectra for treated (top) and untreated (bottom) fique fiber

2.4 Surface Morphology

The effect of alkali concentration on surface morphology was examined using SEM. The effect of fabric softener on as-received fibers before alkalization were also studied. Two different concentrations of NaOH solution were employed in this study. SEM was mainly used for observation of the surface morphology as it provided more details.

2.4.1 Effect of fabric softener. The figure 4 show thee fiber surface of the untreated Sample A is seemingly smooth and intact, although loose helical fibrils can be found occasionally. This could be due to damage during the fiber extraction. The fiber also has a uniform diameter throughout its length. However, the thickness varies between each fiber in the interval 150 – 200 μm. Here it is easier to identify the cells walls. At these magnifications, the surface does not seem very smooth, although the roughness also seems to vary from fiber to fiber. Figure 5 shows that Sample B an increased surface roughness, even more so than the as-received Sample A. The rectangular cell wall edges are slightly more pronounced. One would think the contrary was the case since the individual fiber of Sample E do not adhere to each other. This hints at low mechanical interlocking which should be the case for a smooth surface, which is obviously not the case for Sample B

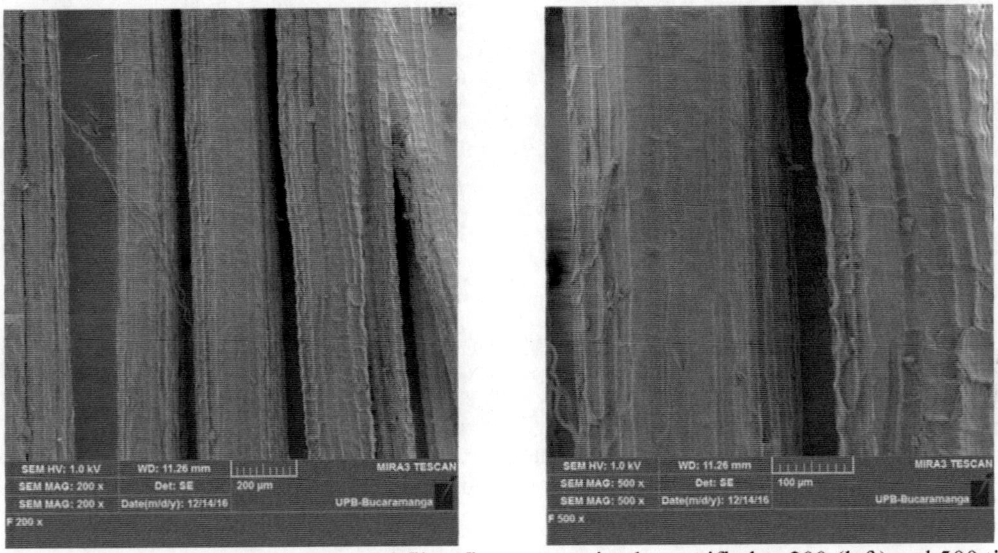

Figure 4. SEM micrographs of untreated fiber fique as received magnified at 200 (left) and 500 times (right)

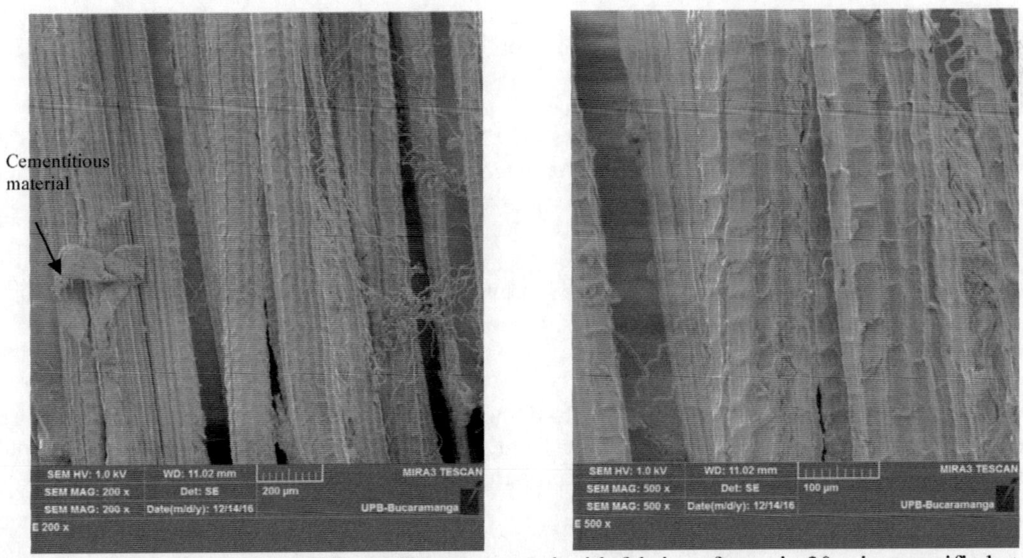

Figure 5. SEM micrographs of fique fiber pretreated with fabric softener in 30 min magnified at 200 (left) and 500 times (right)

The surfaces of pretreated samples appear to be rougher, which is expected as the pretreatment alone had already increased the surface roughness as shown in Figure 6. Like Sample B, the fiber bundles of these pretreated mercerized samples are loosely bound. It is therefore possible that this observation is related to the chemical changes on the surface brought about by the fabric softener. The fiber-fiber adhesion of non-pretreated samples is not solely mechanical but also chemical.

Figure 6. SEM micrographs of pretreated fique fiber after treatment in 2.5 wt% NaOH in 30 min magnified at 200 (left) and 500 times (right)

3. Conclusions

In general, in the TGA analysis of all the treated and non-treated samples it can be noted that it presents two temperatures of weight loss that are appreciable. Making a comparison of the thermograms, it can be noticed that the fique fibers submitted to a previous washing process are thermally less stable that the rest of the fique fibers, remaining more stable the fibers treated previously with NaOH.

The use of alkali solution had successfully changed the surface morphology of the as-received fique fiber as proven by Scanning Electron Microscopy (SEM) and examination of the FTIR spectra. The concentrations 2.5 wt% NaOH resulted in a greater extent of fiber alkalization where helical fibrils were exposed in addition to increased surface roughness.

The use of fabric softener on as-received fibers resulted in increased surface roughness. Despite this fact, there is less mechanical interlocking of individual fibers compared to the as-received samples natural fibers. This property was preserved even after drastic changes on the fiber surface following alkalization.

For future prospects, thermoset and thermoplastic polymers could be used as the composite matrix. Moreover, the several variables such as fiber type, length, orientation, loading and fiber dispersion could be varied to study their effect on the composites properties.

Acknowledgments
This work was supported by the DIT under grant N° 025-0716-2600; this financial support is gratefully acknowledged. The authors would like to express their gratitude to GiDeTechMA Research Group of Faculty Mechanical Engineering, Universidad Pontificia Bolivariana, Bucaramanga.

References

[1] Ministerio de Agricultura y Desarrollo Rural, TECHNIPETROL S.P.A., "Propuesta de Diversificacion industrial de la fibra de Fique," 1986.
[2] C. Garavito, J. Ramírez, C. Garavito, and J. Ramírez, "Cadena de FIQUE y su Agroindustria," 2015.
[3] Cadena Productiva Nacional del fique, *Guía ambiental del subsector fiquero*, Cadena Pro. Colombia, 2006.
[4] Navin Chand and F. Mohammed, "Tribology of natural fiber polymer composites," *Elsevier*, vol. 12, no. 3, p. 45, 2008.
[5] P. Ganan, "Effect of Fiber Treatments on Mechanical Behavior of Short Fique Fiber-reinforced Polyacetal Composites," *J. Compos. Mater.*, vol. 39, no. 7, pp. 633–646, 2005.
[6] P. Gañán and I. Mondragon, "Surface modification of fique fibers. Effect on their physico-mechanical properties," *Polym. Compos.*, vol. 23, no. 3, pp. 383–394, 2002.
[7] P. Gañán and I. Mondragon, "Fique fiber-reinforced polyester composites: Effects of fiber surface treatments on mechanical behavior," *J. Mater. Sci.*, vol. 39, pp. 3121–3128, 2004.
[8] S. Kalia, B. S. Kaith, and I. Kaur, "Pretreatments of natural fibers and their application as reinforcing material in polymer composites-A review," *Polym. Eng. Sci.*, vol. 49, no. 7, pp. 1253–1272, Jul. 2009.
[9] M. Boopalan, M. J. Umapathy, and P. Jenyfer, "A Comparative Study on the Mechanical Properties of Jute and Sisal Fiber Reinforced Polymer Composites," *Silicon*, vol. 4, no. 3, pp. 145–149, 2012.
[10] F. L. Leite, P. S. P. Herrmann, A. L. Da Róz, F. C. Ferreira, A. A. S. Curvelo, and L. H. C. Mattoso, "Investigation of sisal fibers by atomic force microscopy: morphological and adhesive characteristics.," *J. Nanosci. Nanotechnol.*, pp. 2354–2361, 2006.
[11] Y. Li and Y.W. Mai, "Interfacial Characteristics of Sisal Fiber and Polymeric Matrices," *J. Adhes.*, vol. 82, no. 5, pp. 527–554, 2006.
[12] S. Kaewkuk, W. Sutapun, and K. Jarukumjorn, "Effects of interfacial modification and fiber content on physical properties of sisal fiber/polypropylene composites," *Compos. Part B Eng.*, vol. 45, no. 1, pp. 544–549, 2013.
[13] S. L. Fávaro, T. A. Ganzerli, A. G. V de Carvalho Neto, O. R. R. F. da Silva, and E. Radovanovic, "Chemical, morphological and mechanical analysis of sisal fiber-reinforced recycled high-density polyethylene composites," *Express Polym. Lett.*, vol. 4, no. 8, pp. 465–473, 2010.
[14] J. T. Kim and A. N. Netravali, "Mercerization of sisal fibers: Effect of tension on mechanical properties of sisal fiber and fiber-reinforced composites," *Compos. Part A*, vol. 41, pp. 1245–1252, 2010.
[15] S. Panigrahi, "The Effect of Fiber Loading and Chemical Treatment on Mechanical and Thermal Properties of Jute Biocomposites," in *SAE Technical Paper*, 2010.
[16] Q. Zhou, D. Cho, B. K. Song, and H.-J. Kim, "Novel Jute/Polycardanol Biocomposites: Effect of Fiber Surface Treatment on Their Properties," *Compos. Interfaces*, vol. 16, no. 7–9, pp. 781–795, 2009.
[17] A. S. Ahmed, M. S. Islam, A. Hassan, M. K. Mohamad Haafiz, K. N. Islam, and R. Arjmandi, "Impact of succinic anhydride on the properties of jute fiber/polypropylene biocomposites," *Fibers Polym.*, vol. 15, no. 2, 2014.
[18] J. A. Khan, M. A. Khan, and R. Islam, "Effect of mercerization on mechanical, thermal and degradation characteristics of jute fabric-reinforced polypropylene composites," *Fibers Polym.*, vol. 13, no. 10, 2012.
[19] N. Gibeop, D. W. Lee, C. V. Prasad, F. Toru, B. S. Kim, and J. I. Song, "Effect of plasma treatment on mechanical properties of jute fiber/poly (lactic acid) biodegradable composites," *Adv. Compos. Mater.*, vol. 22, no. 6, 2013.
[20] H. Jena, M. K. Pandit, and A. K. Pradhan, "Effect of cenosphere on mechanical properties of

bamboo-epoxy composites," *J. Reinf. Plast. Compos.*, vol. 32, no. 11, pp. 794–801, 2013.
[21] F. He, C. Gao, and S. Ye, "Effect of Bamboo Fiber Modification on Tribological Performance of Brake Composites," *Adv. Mater. Res.*, vol. 151–151, pp. 1801–1805, 2011.
[22] H. P. S. A. Khalil, I. U. H. Bhat, M. Jawaid, A. Zaidon, D. Hermawan, and Y. S. Hadi, "Bamboo fibre reinforced biocomposites: A review," *Mater. Desing*, vol. 42, pp. 353–368, 2012.
[23] D. Liu, J. Song, D. P. Anderson, P. R. Chang, and Y. Hua, "Bamboo fiber and its reinforced composites: Structure and properties," *Cellulose*, vol. 19, no. 5. pp. 1449–1480, Oct-2012.
[24] N. Sarifuddin and H. Ismail, "Applications of Kenaf-Lignocellulosic Fiber in Polymer Blends," *Lignocellul. Polym. Compos. Process. Charact. Prop.*, pp. 499–521, 2015.
[25] Z. N. Azwa and B. F. Yousif, "Characteristics of kenaf fibre/epoxy composites subjected to thermal degradation," *Polym. Degrad. Stab.*, vol. 98, no. 12, pp. 2752–2759, 2013.
[26] W. M. Z. W. M. Z. W. Yunus *et al.*, "Effect of chemical surface treatment on the mechanical properties of reinforced plasticized poly(lactic acid) biodegradable composites," *J. Reinf. Plast. Compos.*, vol. 30, no. 5, pp. 381–388, 2011.
[27] H. Anuar and A. Zuraida, "Improvement in mechanical properties of reinforced thermoplastic elastomer composite with kenaf bast fibre," *Compos. Part B Eng.*, vol. 42, no. 3, pp. 462–465, 2011.
[28] H. Hajiha, M. Sain, and L. H. Mei, "Modification and Characterization of Hemp and Sisal Fibers," *J. Nat. Fibers*, vol. 11, no. 2, pp. 144–168, 2014.
[29] S. H. Aziz and M. P. Ansell, "The effect of alkalization and fibre alignment on the mechanical and thermal properties of kenaf and hemp bast fibre composites: Part 1 – polyester resin matrix," 2003.
[30] A. K. . G. J. Bledzki, "Composites reinforced with cellulose based fibers.," *1999*, vol. 24, pp. 221–274, 1999.
[31] C. Gómez Hoyos, V. A. Alvarez, P. G. Rojo, and A. Vázquez, "Fique fibers: Enhancement of the tensile strength of alkali treated fibers during tensile load application," *Fibers Polym.*, vol. 13, no. 5, pp. 632–640, 2012.
[32] Y. Y. Then, N. A. Ibrahim, N. Zainuddin, B. W. Chieng, H. Ariffin, and W. M. Z. Wan Yunus, "Effect of 3-Aminopropyltrimethoxysilane on Chemically Modified Oil Palm Mesocarp Fiber/Poly(butylene succinate) Biocomposite," *BioResources*, vol. 10, no. 2, pp. 3577–3601, Apr. 2015.
[33] Y. Xie, C. A. S. Hill, Z. Xiao, H. Militz, and C. Mai, "Silane coupling agents used for natural fiber/polymer composites: A review," *Compos. Part A Appl. Sci. Manuf.*, vol. 41, no. 7, pp. 806–819, 2010.
[34] F. L. Vera, H. A. M. Cortes, C. V. Murcia, and I. C. Galvis, "Modificación superficial de micro fibras de celulosa obtenidas a partir de bagazo de caña de azúcar usando silanización," *Rev. Inf. Técnico*, vol. 78, no. 2, pp. 106–114, 2014.
[35] D. Cho, H. S. Lee, and S. O. Han, "Effect of Fiber Surface Modification on the Interfacial and Mechanical Properties of Kenaf Fiber-Reinforced Thermoplastic and Thermosetting Polymer Composites," *Compos. Interfaces*, vol. 16, no. January 2013, pp. 711–729, 2009.
[36] M. S. Huda, L. T. Drzal, A. K. Mohanty, and M. Misra, "Effect of chemical modifications of the pineapple leaf fiber surfaces on the interfacial and mechanical properties of laminated biocomposites," *Compos. Interfaces*, vol. 15, no. 2–3, pp. 169–191, 2008.
[37] J. Zhu, H. Zhu, J. Njuguna, and H. Abhyankar, "Recent development of flax fibres and their reinforced composites based on different polymeric matrices," *Materials (Basel).*, vol. 6, no. 11, pp. 5171–5198, 2013.
[38] H. U. Zaman and M. Beg, "Preparation, structure, and properties of the coir fiber/polypropylene composites," *J. Compos. Mater.*, vol. 48, no. 26, pp. 3293–3301, 2014.
[39] Y. Xie, C. A. S. Hill, Z. Xiao, H. Militz, and C. Mai, "Silane coupling agents used for natural fiber/polymer composites: A review," *Compos. Part A Appl. Sci. Manuf.*, vol. 41, no. 7, pp.

806–819, Jul. 2010.
[40] E. Corradini, L. de Morais, M. de F. Rosa, S. E. Mazzetto, L. H. C. Mattoso, and J. A. M. Agnelli, "A preliminary study for the use of natural fibers as reinforcement in starch-gluten-glycerol matrix," *Macromol. Symp.*, vol. 245, pp. 558–564, 2006.
[41] C. Castro, A. Palencia, I. Gutiérrez, G. Vargas, and P. Gañán, "Determination of optimal alkaline treatment conditions for fique fiber bundles as reinforcement of composites materials," *Rev. Téc. Ing. Univ. Zulia*, vol. 30, no. 2, pp. 136–142, 2007.
[42] P. Gañán and I. Mondragon, "Influence of Compatibilization Treatments on the Mechanical Properties of Fique Fiber Reinforced Polypropylene Composites," *Int. J. Polym. Mater.*, vol. 53, no. 11, pp. 997–1013, 2004.
[43] A. Shalwan and B. F. Yousif, "In state of art: Mechanical and tribological behaviour of polymeric composites based on natural fibres," *Mater. Des.*, vol. 48, pp. 14–24, 2013.
[44] M. A. Hidalgo-Salazar, J. H. Mina, and P. J. Herrera-Franco, "The effect of interfacial adhesion on the creep behaviour of LDPE-Al-Fique composite materials," *Compos. Part B Eng.*, vol. 55, pp. 345–351, 2013.
[45] H. M. Ng *et al.*, "Extraction of cellulose nanocrystals from plant sources for application as reinforcing agent in polymers," *Compos. Part B Eng.*, vol. 75, pp. 176–200, 2015.
[46] A. Mtibe, Y. Mandlevu, L. Z. Linganiso, and R. D. Anandjiwala, "Extraction of cellulose nanowhiskers from flax fibres and their reinforcing effect on poly(furfuryl) alcohol," *J. Biobased Mater. Bioenergy*, vol. 9, no. 3, 2015.
[47] M. Asim, M. Jawaid, K. Abdan, and M. R. Ishak, "Effect of Alkali and Silane Treatments on Mechanical and Fibre-matrix Bond Strength of Kenaf and Pineapple Leaf Fibres," *J. Bionic Eng.*, vol. 13, no. 3, pp. 426–435, 2016.
[48] A. Amiri, C. A. Ulven, and S. Huo, "Effect of chemical treatment of flax fiber and resin manipulation on service life of their Composites using time-temperature superposition," *Polymers (Basel).*, vol. 7, no. 10, pp. 1965–1978, 2015.
[49] A. C. Manalo, E. Wani, N. A. Zukarnain, W. Karunasena, and K. Lau, "Effects of alkali treatment and elevated temperature on the mechanical properties of bamboo fibre–polyester composites," *Compos. Part B Eng.*, vol. 80, no. 0, pp. 73–83, 2015.
[50] J. A. Khan, M. A. Khan, and R. Islam, "Effect of Mercerization on Mechanical, Thermal and Degradation Characteristics of Jute Fabric-reinforced Polypropylene Composites," *Fibers Polym.*, vol. 13, no. 10, pp. 1300–1309, 2012.
[51] S. A. Paul, A. Boudenne, L. Ibos, Y. Candau, K. Joseph, and S. Thomas, "Effect of fiber loading and chemical treatments on thermophysical properties of banana fiber/polypropylene commingled composite materials."
[52] F. He, C. Gao, and S. Ye, "Effect of Bamboo Fiber Modification on Tribological Performance of Brake Composites."
[53] A. Atiqah, M. A. Maleque, M. Jawaid, and M. Iqbal, "Development of kenaf-glass reinforced unsaturated polyester hybrid composite for structural applications," *Compos. Part B*, vol. 56, pp. 68–73, 2014.
[54] P. Ferreira De Oliveira, M. De Fatima, and V. Marques, "Chemical treatment of natural Malva fibres and preparation of green composites with Poly(3-Hydroxybutyrate)," *Chem. Chem. Technol.*, vol. 9, no. 2, 2015.
[55] K. Nobuta *et al.*, "Characterization of cellulose nanofiber sheets from different refining processes," *Cellulose*, vol. 23, no. 1, 2016.
[56] S. Lee, H. Pan, C. Y. Hse, A. R. Gunasekaran, and T. F. Shupe, "Characteristics of regenerated nanocellulosic fibers from cellulose dissolution in aqueous solutions for wood fiber/polypropylene composites," *J. Thermoplast. Compos. Mater.*, vol. 27, no. 4, pp. 558–570, 2014.
[57] S. Zarina and I. Ahmad, "Biodegradable Composite Films based on κ-carrageenan Reinforced by Cellulose Nanocrystal from Kenaf Fibers," *BioResources*, vol. 10, no. 1, pp. 256–271, Nov.

IOP Conf. Series: Materials Science and Engineering **437** (2018) 012015 doi:10.1088/1757-899X/437/1/012015

2014.

[58] O. Faruk, A. K. Bledzki, H.-P. Fink, and M. Sain, "Biocomposites reinforced with natural fibers: 2000–2010," *Prog. Polym. Sci.*, vol. 37, no. 37, pp. 1552–1596, 2012.

[59] L. Zhao, W. Xia, K. Tarverdi, and J. Song, "Biocomposite boards from wheat straw without addition of bonding agent," *Mater. Sci. Technol. (United Kingdom)*, vol. 30, no. 5, 2014.

[60] M. Boopalan, M. J. Umapathy, and P. Jenyfer, "A Comparative Study on the Mechanical Properties of Jute and Sisal Fiber Reinforced Polymer Composites," *Silicon*, vol. 4, no. 3, pp. 145–149, Jul. 2012.

[61] S. Siyamak, N. A. Ibrahim, S. Abdolmohammadi, W. M. Z. B. W. Yunus, and M. Z. A. B. Rahman, "Enhancement of mechanical and thermal properties of oil palm empty fruit bunch fiber poly(butylene adipate-co-terephtalate) biocomposites by matrix esterification using succinic anhydride," *Molecules*, vol. 17, no. 2, pp. 1969–1991, 2012.

[62] S. Siyamak, N. A. Ibrahim, S. Abdolmohammadi, W. M. Z. B. W. Yunus, and M. Z. A. Rahman, "Enhancement of Mechanical and Thermal Properties of Oil Palm Empty Fruit Bunch Fiber Poly(butylene adipate-co-terephtalate) Biocomposites by Matrix Esterification Using Succinic Anhydride," *Molecules*, vol. 17, no. 12, pp. 1969–1991, Feb. 2012.

[63] P. G. A. Restrepo, C. Bautista, "Influencia de modificaciones tipo esterificación, sobre el comportamiento hidrofílico de reforzantes celulósicos," *Investig. Apl.*, vol. numero 3, no. 2008, p. 9, 2008.

[64] P. Gañán and I. Mondragon, "Thermal and degradation behavior of fique fiber reinforced thermoplastic matrix composites," *J. Therm. Anal. Calorim.*, vol. 73, no. 3, pp. 783–795, 2003.

[65] S. Qian, H. Mao, E. Zarei, and K. Sheng, "Preparation and Characterization of Maleic Anhydride Compatibilized Poly(lactic acid)/Bamboo Particles Biocomposites," *J. Polym. Environ.*, vol. 23, no. 3, pp. 341–347, 2015.

[66] S. Mishra, J. B. Naik, and Y. P. Patil, "The compatibilising effect of maleic anhydride on swelling and mechanical properties of plant-fiber-reinforced novolac composites," *Compos. Sci. Technol.*, vol. 60, pp. 1729–1735, 2000.

[67] A. K. Rana, A. S. Singha, M. Kumari, and V. Kumar, "Effect of Chemical Functionalization on Functional Properties of Cellulosic Fiber-Reinforced Polymer Composites," *Lignocellul. Polym. Compos.*, pp. 281–99, 2014.

[68] R. Campilho, "Introduction to Natural Fiber Composites," in *Natural Fiber Composites*, CRC Press, 2015, pp. 1–34.

[69] S. N. Monteiro, V. Calado, R. J. S. Rodriguez, and F. M. Margem, "Thermogravimetric stability of polymer composites reinforced with less common lignocellulosic fibers - An overview," *J. Mater. Res. Technol.*, vol. 1, no. 2, pp. 117–126, 2012.

IOP Conf. Series: Materials Science and Engineering **437** (2018) 012016 doi:10.1088/1757-899X/437/1/012016

Synthesis and morphological characterization of Li-Ti/PVP fibers as precursors for $Li_4Ti_5O_{12}$ towards its future use as anode materials in Li-ion fiber batteries by means of Electrospinning

N Castano[1,2], M A Cortes[1,2], E Garcia[1,2,3] and H V Martinez[1,2]

[1] Mechanical Engineering Department. Grupo de investigación sobre Nuevos Materiales (GINUMA), Universidad Pontificia Bolivariana, Medellín, Colombia
[2] Mechanical Engineering Department. Grupo de investigación Energía y Termodinámica (GET), Universidad Pontificia Bolivariana, Medellín, Colombia

[3] Nanotechnology Engineering Department, Universidad Pontificia Bolivariana, Medellín, Colombia

nicolas.castano@upb.edu.co

Abstract. The ever-increasing evolution and popularity of electronic devices, in hand with globalization, has led to a highly innovative market, always in movement. Novel applications such as smart textiles, where portable electronics are coupled to fabrics, are expected to become a worldwide trend when they overcome some limitations, especially with the energy storage systems used to power them. It is in this context where a flexible battery with the shape of a fiber is of importance. Electrospinning is a versatile synthesis technique where micro and nano fibers can be obtained. When these fibers are conducted onto a substrate it is possible to produce materials with different characteristics and morphologies. In this work, Li4Ti5O12 (LTO) was synthesized as anode material for li-ion batteries and was further characterized. LTO is attractive as it presents characteristics such as high thermal stability, relatively high volumetric capacity and high cyclability. The synthesis of this material was developed in two steps. First, a precursor solution containing a spinnable polymer, and titanium and lithium salts, which were dissolved in a mixture of solvents, was subjected to a process of electrospinning. Then, the obtained fibers were calcined at temperatures between 650 and 850 °C for 7-10 hours in air or argon. Scanning electron microscopy (SEM) coupled to energy dispersive x-ray spectroscopy (EDS) was used to study its morphology and elemental composition. Meanwhile, thermogravimetric analysis (TGA) was performed to study the thermal stability.

1. Introduction

The ever-increasing degree of portability in electronic devices, where there is a constant requirement for reduction in size and weight, makes evident the need for more efficient energy storage systems to power these applications. Likewise, hyper connectivity in today's society is not something temporary but, on the contrary, there is a worldwide trend moving towards an inter-networking of physical devices, vehicles and even buildings, which has been denoted as Internet of Things (IoT) [1][2]. This game changing idea calls for objects to collect and exchange data with the possibility that, using a cellphone, different electronic devices can be controlled [1][3], i.e. physiological vital signs from a patient [4][5][6] or electrocardiogram signals from animals by means of Smart textiles [7][8] could be remotely monitored. The design of such monitoring systems is more common every day due to their potential application in medicine, sports and security [9]. Nevertheless, the performance and massive market penetration of these type of technologies depend critically on the development of energy storage systems with adequate configuration and geometry, which can be adapted for different applications [10][11][12].

In this regard, it is crucial to develop better materials to increase the efficiency of energy storage systems [13]. In particular, nanotechnology is one of the strategies people use for the improvement of materials for anodes, cathodes, electrolytes and other components, as the increase in surface area provided when dealing with materials at this scale offers more reaction and exchange sites [14], which allows for the design of electrodes with higher energy and power densities and stability upon cycling [15].

In Li-ion batteries, graphite is the material typically used for the anode but it has issues associated with mechanical stability upon cycling, operative safety and relatively slow kinetics during charge/discharge [16]. Different alternatives for graphite as anode material have been investigated. Among them, the spinel material Li4Ti5O4 (LTO) stands out in terms of stability upon successive lithium ion insertion/removal. It has a decent theoretical capacity of 175 mAh/g, at a voltage of 1,55 V vs Li+/Li [17] and is dubbed as a “zero stress” material as the phase change during lithiation/ delithiation results in very small volumetric changes (≈0,2%) [18]. This feature makes it a material with an increased safety (it suppresses the formation of the solid electrolyte interphase -SEI- and impedes the formation of lithium dendrites), highly durable as it presents a good cycling life cycling rate and it is able to withstand high charge/discharge current rates [19][20][21]. It is therefore a strong candidate to circumvent several technical issues associated with batteries, both in stationary electricity storage and portable applications [22].

Electrospinning is a synthesis technique which permits the creation of fibers from a polymer solution. In it, a precursor solution is loaded into a syringe and is subjected to a high voltage and, depending on operation parameters of both the equipment (voltage, flow rate, distance between tip and collector) and the solution (concentration, viscosity, surface tension, conductivity), the formation of fibers, sometimes with diameters in the nano-range, can be achieved. These geometrical morphology could open doors in the development of flexible batteries with potential application in smart textiles [23][24][25].

This article describes the procedure to obtain a working precursor solution formulation for the production of Li4Ti5O12 fibers to be used as anode material for Li-ion fiber batteries.

2. Materials and Methods

2.1. Preparation of precursor solution.

The process of preparation of a working precursor solution was iterative and was initially based on formulations found in the literature [26],[27]. To obtain Li4Ti5O12 fibers, precursors for each component (i.e., lithium, titanium and carbon) were needed.

Following the formulation suggested by Bingkun Guo et al [28], isopropyl alcohol was used as solvent for polivinilpyrrolidone (PVP) as well as for lithium acetate and titanium butoxide precursor salts. However, the solution was not stable as it precipitated during the electrospinning process. On the other hand, other authors suggested a mixture of solvents such as acetic acid, ethanol, acetone and acetylacetone [29],[30]. Following this formulation, a stable solution was obtained for the timespan of the electrospinning process and, thus, two precursor solutions were prepared: in solution 1 (Sln 1), 7.5 g of ethanol, 1.5 g of acetic acid and 2.5 g of acetone were mixed during 10 minutes under constant stirring. Then, 0.5 g of lithium acetate, 3 g of titanium butoxide and 5 g of polivinilpyrrolidone (PVP, MW: 55000) were added under constant stirring during 24 h at 50 ° C. Solution 2 (Sln 2) was prepared by mixing 9 g of ethanol, 2.5 g of acetic acid under constant stirring for 10 minutes, followed by an addition of 0.5 g of lithium acetate and 3 g of titanium butoxide under constant stirring during 60 minutes. Finally, 5 g of polivinilpyrrolidone (PVP, MW: 55000) was added under constant stirring during 24 h at 50 °C. All chemicals used were purchased from Sigma-Aldrich.

IOP Conf. Series: Materials Science and Engineering **437** (2018) 012016 doi:10.1088/1757-899X/437/1/012016

2.2. Rheological characterization
The rheology of the precursor solutions described above, was analyzed using a DHR2 rheometer (Discovery Hybrid Rheometers) with a standard 25 mm diameter stainless steel flat plate geometry, 50.0 μm gap between plates, a tension of 2%, with an applied oscillatory strain of 10 Hz and temperature control at 25 °C. Pretreatment curves were done at a shearing rate of 10 1/s during 10 seconds, followed by flux scan in the range between 0.1-200 1/s point by point, with an average value of 10 s per point.

2.3. Thermal characterization
Thermogravimmetric analysis (TGA) was done using a thermobalance (Mettler Toledo TGA/SDTA 851 E) between 30 - 800 °C with a heating rate of 10 °C/min under nitrogen. Two analyses were performed, one with PVP to have a comparison (blank) and the second with the Li-Ti/PVP precursor fibers.

2.4. Electrospinning setup
A schematic drawing of the setup used for the production of Li-Ti/PVP fibers is shown un figure 1 (right). It consists in a modular system with three main elements: (1) a high voltage source (5-30 kV); (2) a syringe pump, to deliver solutions in a controlled way (flow rate and volume); and (3) a rotating drum which acts as collector and that is grounded to close the circuit. Additional accessories such as (4) and (5) are a ventilation system and a high intensity lamp, respectively.

When the precursor solution is flowing at a specific flow rate, a droplet starts forming at the tip of the nozzle. When it is subjected to a high voltage, the forces acting in the droplet reach a point in which the so-called Taylor cone [31] is formed, mainly due to electrostatic effects, and a jet emerges from the apex of the cone. Inner forces such as surface tension prevent the breakage of the droplet and, thus, the jet becomes elongated forming a fiber. Depending on operation parameters of both the equipment (voltage, flow rate, distance between tip and collector) and the solution (concentration, viscosity, surface tension, conductivity), the formation of fibers with diameters in the micro/nano-range, can be achi

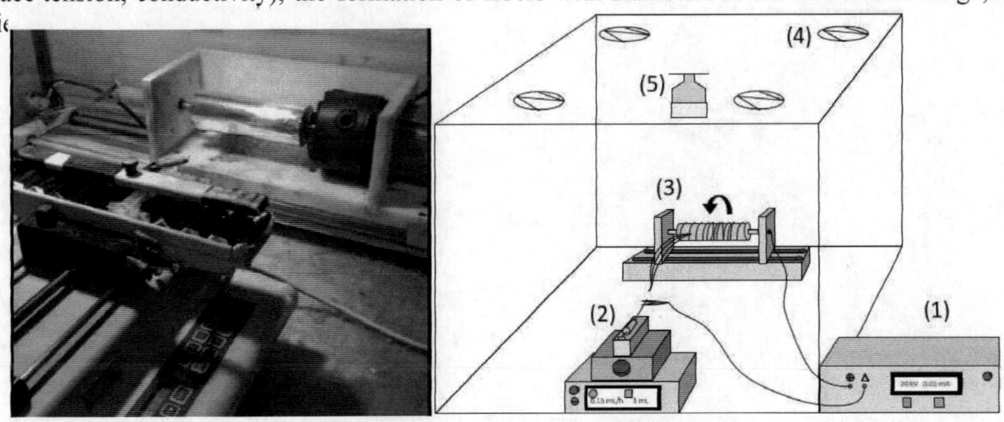

Figure 1. left, electrospinning photograph and right, schematic drawing showing the basic elements comprising this technique

As it can be appreciated in figure 1, electrospinning is a simple and versatile technique, that is relatively easy to setup, which permits the production of fibers at the micro/nano scale for a large variety of applications, due in part for the wide range of polymers that can be used [32][33]. The morphology of the materials produced with this technique can be varied by tuning the parameters associated with the precursor solution (viscosity, concentration, molecular weight of the polymer, surface tension, conductivity, solvent type), the setup (voltage, flow rate, distance from tip to substrate, geometry of the collector) and the environmental conditions (temperature and humidity) [34][35].

In this research, the setup parameters were fixed while the behavior of precursor solutions 1 and 2 were evaluated. Thus, both solutions were subjected to a voltage of around 20 kV (Gamma RS232 power source), a flow rate of 0.2 mL/h and a distance of 15 cm from tip to a rotating cylindrical drum covered with aluminum foil.

2.5. Thermal treatment.

The obtained electrospun fibers were pre-dried between 80-90 °C during 30 minutes. Then, the deposited film was scratched from the collector surface and was calcined at 750°C during 10 h in air, following similar reports in the literature [36].

2.6. Chemical and morphological characterization.

The morphology, size and size distribution of samples were evaluated using a scanning electron microscope (SEM, JEOL JCM-6000 Plus) coupled with an energy dispersive x-spectroscopy (EDS) detector for chemical composition analysis.

3. Results and Discussions

3.1. Rheological characterization.

The rheology of the precursor solutions is shown in figure 2. Both present a behavior similar to a dilatant fluid (i.e. a liquid that presents an increase in viscosity with the increase of shear rate) on the first points of the curve [37]. For Sln 1, this behavior goes until a shear rate of 20 1/s and for Sln 2 until 46.5 1/s.

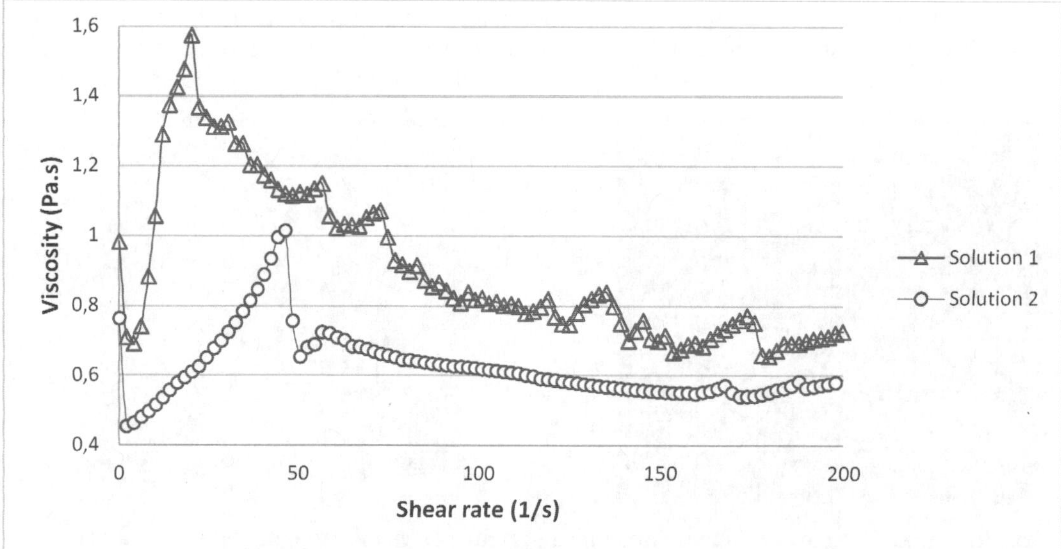

Figure 2. Rheological behavior of the precursor solutions

After these inflection points, the behavior of Sln 1 tends towards a pseudo plastic liquid [38], while Sln 2 looks more like a Newtonian fluid with a viscosity of ~0.6 Pa.s. According to the dilatant behavior in both solutions, it seems convenient to operate the electrospinning at high voltages to guarantee a small range of viscosities during the process, as larger viscous forces affect the size of the fibers [39][40]. It is worth mentioning that to know the exact operating conditions during electrospinning, a correlation

IOP Conf. Series: Materials Science and Engineering **437** (2018) 012016 doi:10.1088/1757-899X/437/1/012016

between the shear rate and applied voltage should be considered. In this regard, it is known that for a liquid flowing through a cylindrical cavity there is a proportional relation between shear rate and flow rate ([shear rate] α [flow rate]). Also, it's been experimentally observed that the diameter of the jet decreases with the increase on voltage. However, this correlation is not completely understood at present and is beyond the scope of this work.

3.2. Thermal characterization:

Thermogravimmetric analysis (fig. 3) of both the precursor fibers obtained through electrospinning of Sln 2 and the polymer used as precursor (PVP, MW: 55000).

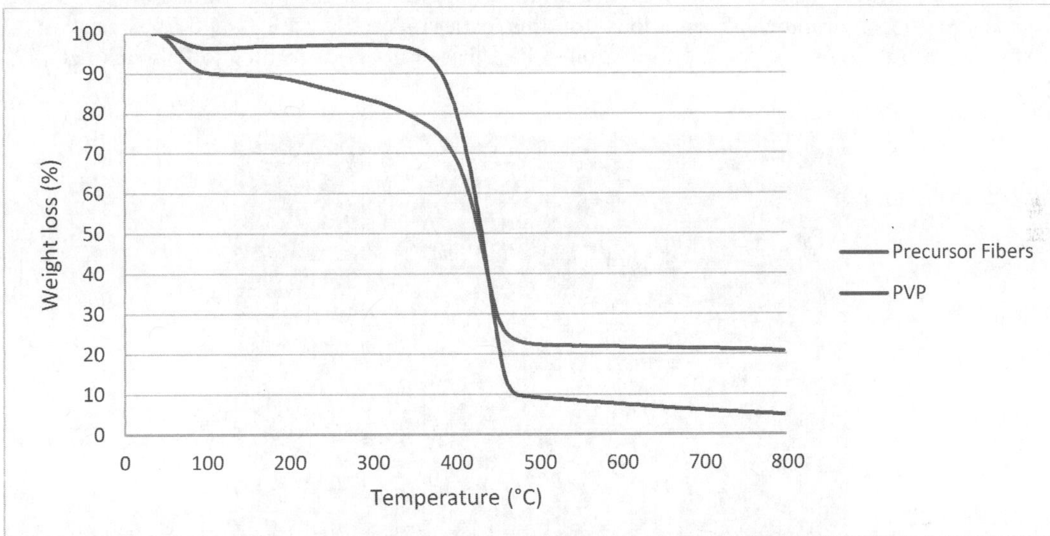

Figure 3. Thermogravimmetric analysis of PVP and as-obtained Li-Ti/PVP fibers

A summary of the relevant thermal events present in the thermogram are shown in table 1. It is observed that for the precursor fibers there is loss of humidity between 46 and 107°C, corresponding to a 10.1% loss of mass. For the PVP, there is loss of humidity between 46 and 104°C, corresponding to 3.7% of its mass.

Table 1. Summary of thermal events

	Mass (mg)	Thermal event	T_i (°C)	T_f (°C)	$\%M_i$	$\%M_f$
		1	46	104.2	100	96.3
PVP	8.8423	2	335.2	477.1	96.3	9.6
		3	477.1	780	9.6	5
		1	46	107.1	100	89.9
Precursor Fibers	8.2322	2	135	500.7	89.8	22.2
		3	500.7	780	22.2	20.8

In the second event for the PVP, between 335 and 477°C most of the carbon is eliminated, corresponding to 90% of its mass. Likewise, for the Li-Ti/PVP fibers, from 135 and 500°C carbon is eliminated, corresponding to 75% of its mass. After this, the mass of the precursor fibers remains almost constant suggesting that the reaction is over. Therefore, to obtain the Li4Ti5O12 composite a calcination temperature above 500°C is required. Similar reports in the literature suggest a range of temperature between 500-780°C [29][41].

3.3. Morphological and chemical characterization

Several attempts were made with Sln 1 to obtain fibers. Figure 4 shows the outcome of most of them and it can be clearly seen that no fibers were formed by using this solution. This result looks closer to an electrospray process where instead of fibers there is a generation of droplets or particles [42]. The average diameter of the obtained particles is 3.1 μm with a standard deviation of 1.0 μm. The main difference between solutions 1 and 2 are the mix of solvents used to dissolve the precursor salts and the polymer. This sole variation can change properties such as surface tension, boiling point and rheological behavior. The boiling point of the precursor solution is crucial as the solvent evaporation of the formed jet during the time of flight it takes to travel between the tip of the nozzle and the collector drum clearly defines the final morphology [34] [43]. Besides, the environmental conditions for this experiment, i.e 24.3°C and 70% relative humidity, were not favorable for the formation of PVP-based fibers, according to Pelipenko et al. [44].

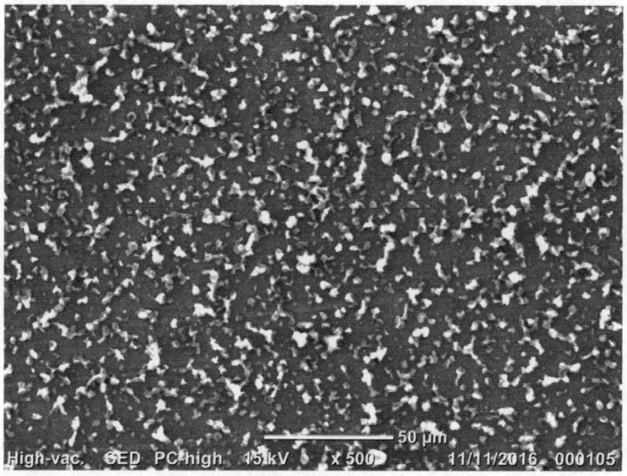

Figure 4. SEM micrograph of an electrospinning process with Sln 1(with acetone)

In figure 5, a set of micrographs of fibers obtained with Sln 2 are shown. Fig. 5A-B represent the Li-Ti/PVP precursor fibers, as-obtained with electrospinning, while fig. 5C-D represent the LTO fibers after calcination at 750°C in 10 hours in air.

The average diameter of Li-Ti/PVP fibers is 0.87 μm with a standard deviation of 0.63 μm, while the LTO fibers have a mean diameter of 1.23 μm and a standard deviation of 0.74 μm. The parameters used to obtain these fibers were: flow rate, 0.2 mL/h; voltage, 20 kV; distance, 15 cm; 19 G nozzle (external diameter, 1.07 mm and internal diameter, 0.69 mm); collector drum rotation speed, 20 rpm-clockwise; temperature, 26-27°C; and humidity, 40-50% (these environmental conditions were more favorable for the production of fibers [42][45]).

EDS spectrum of the LTO fibers (after calcination) are shown in figure 6. It reveals the presence of Ti and O, both of them crucial to form the Li4Ti5O12 stoichiometry and the spinel structure [46]. Quantitative analyses gave a 49.60% mass percentage and 74.66% atomic percentage for oxygen, while for titanium it gave 50.40% mass and 25.34% atomic. These results suggest a molar relation of 1:2.95 titanium to oxygen, respectively, and are very promising when looking at the molar ratio for LTO of 1:2.4 [47].

IOP Conf. Series: Materials Science and Engineering **437** (2018) 012016 doi:10.1088/1757-899X/437/1/012016

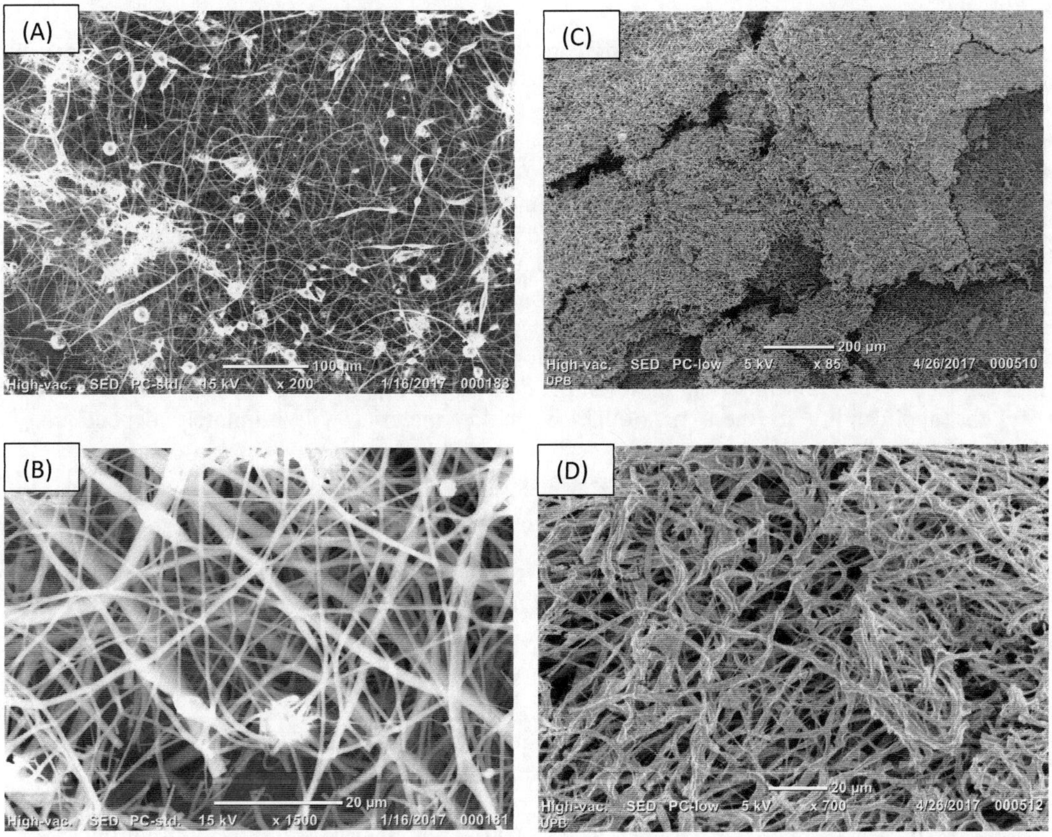

Figure 5. SEM images. (A-B) Li-Ti/PVP precursor fibers before calcination. (C-D) LTO fibers after calcination at 750°C for 10 h in air

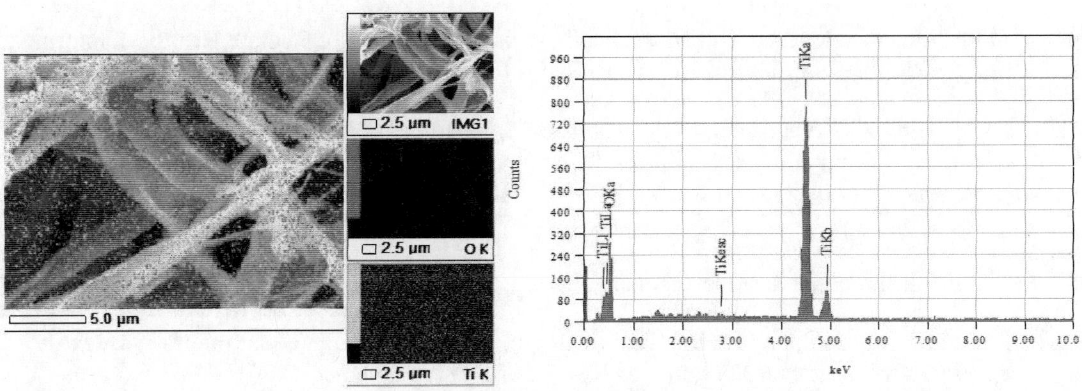

Figure 6. Energy-dispersive X-ray spectroscopy (EDS) of electrospun LTO fibers (0.2 mL/h, 20 kV, 15 cm)

4. Conclusions
In this study, nanofibers based on lithium and titanium salts, together with PVP polymer were successfully produced by means of electrospinning. Moreover, upon calcination at 750°C, after finding a temperature window for this process using thermogravimetric analysis, the fibers maintained their morphology. After chemical composition analysis, the atomic percentage of oxygen to titanium in these fibers matched that of the $Li_4Ti_5O_{12}$ spinel.

Two different solutions were used, employing different solvent mixtures, while keeping all other parameters involved in electrospinning fixed. Fibers were not obtained using solution 1 (with acetone), contrary to solution 2, with the same equipment parameters (voltage, flow rate, distance). It is important to clarify that this electrospinning parameters are recommended for the specific precursors used here, since properties of the solution such as concentration, viscosity, surface tension and conductivity will vary and hence, the optimal electrospinning parameters will also change.

It is relevant to mention that this is a preliminary step on the synthesis of full LTO anode fibers. Finally, the applicability of these nanofibers as anode material will ultimately depend on their electrochemical performance.

This is an ongoing project and further work is needed.

Acknowledgments
The work carried out in this project has been financed by Universidad Pontificia Bolivariana and Patrimonio Autónomo Fondo Nacional de Financiamiento para las Ciencias, la Tecnología y la Innovación, Francisco José de Caldas.

References
[1] T. Saarikko, U. H. Westergren, and T. Blomquist, "The Internet of Things: Are you ready for what's coming?," *Bus. Horiz.*, vol. 60, no. 5, pp. 667–676, 2017.

[2] L. Atzori, A. Iera, and G. Morabito, "Understanding the Internet of Things: definition, potentials, and societal role of a fast evolving paradigm," *Ad Hoc Networks*, vol. 56, pp. 122–140, 2017.

[3] J. M. Jornet and I. F. Akyildiz, "The Internet of Multimedia Nano-Things," *Nano Commun. Netw.*, vol. 3, no. 4, pp. 242–251, Dec. 2012.

[4] S. Majumder, T. Mondal, and M. J. Deen, "Wearable Sensors for Remote Health Monitoring.," *Sensors (Basel).*, vol. 17, no. 1, Jan. 2017.

[5] G. Appelboom, E. Camacho, M. E. Abraham, S. S. Bruce, E. L. Dumont, B. E. Zacharia, R. D'Amico, J. Slomian, J. Y. Reginster, O. Bruyère, E. S. Connolly, and Jr, "Smart wearable body sensors for patient self-assessment and monitoring.," *Arch. Public Health*, vol. 72, no. 1, p. 28, 2014.

[6] "Biomedical monitor for smartphone," Dec. 2012.

[7] A. Guidi, A. Lanata, G. Valenza, E. P. Scilingo, and P. Baragli, "Validation of smart textile electrodes for electrocardiogram monitoring in free-moving horses," 2017.

[8] V. Koncar, "1 – Introduction to smart textiles and their applications," in *Smart Textiles and their Applications*, 2016, pp. 1–8.

[9] T. Yilmaz, R. Foster, and Y. Hao, "Detecting vital signs with wearable wireless sensors.," *Sensors (Basel).*, vol. 10, no. 12, pp. 10837–62, 2010.

[10] C. Wang and G. G. Wallace, "Flexible Electrodes and Electrolytes for Energy Storage," *Electrochim. Acta*, vol. 175, pp. 87–95, Sep. 2015.

[11] D. Miorandi, S. Sicari, F. De Pellegrini, and I. Chlamtac, "Internet of things: Vision, applications and research challenges," *Ad Hoc Networks*, vol. 10, no. 7, pp. 1497–1516, Sep. 2012.

[12] V. L. Pushparaj, M. M. Shaijumon, A. Kumar, S. Murugesan, L. Ci, R. Vajtai, R. J. Linhardt, O. Nalamasu, and P. M. Ajayan, "Flexible energy storage devices based on nanocomposite paper.," *Proc. Natl. Acad. Sci. U. S. A.*, vol. 104, no. 34, pp. 13574–7, Aug. 2007.

[13] C. J. Brinker and D. Ginger, "Nanotechnology for Sustainability: Energy Conversion, Storage, and Conservation," in *Nanotechnology Research Directions for Societal Needs in 2020*, Dordrecht: Springer Netherlands, 2011, pp. 261–303.

[14] A. K. Hussein, "Applications of nanotechnology in renewable energies—A comprehensive overview and understanding," *Renew. Sustain. Energy Rev.*, vol. 42, pp. 460–476, 2015.

[15] G. Zhou, F. Li, H.-M. Cheng, L. Gao, W. P. Tang, J. Y. Xie, J. Zhi, X. M. Ju, C. Scherr, S. Kaskel, C. Wang, L. Hu, C. J. Yu, C. Chuang, T. I. Kim, T. Song, K. Shigeta, S. Kang, C. Dagdeviren, I. Petrov, P. V. Braun, Y. G. Huang, U. Paik, and J. A. Rogers, "Progress in flexible lithium batteries and future prospects," *Energy Environ. Sci.*, vol. 7, no. 4, pp. 1307–1338, Mar. 2014.

[16] X. Sun, P. V. Radovanovic, B. Cui, C. H. Chen, X. W. Ge, V. Gariepy, K. Galoutov, P. Hovington, A. Mauger, H. Groult, and C. M. Julien, "Advances in spinel Li4Ti5O12 anode materials for lithium-ion batteries," *New J. Chem.*, vol. 39, no. 1, pp. 38–63, Dec. 2015.

[17] L. Wang, Q. Xiao, Z. Li, G. Lei, P. Zhang, and L. Wu, "Synthesis of Li4Ti5O12 fibers as a high-rate electrode material for lithium-ion batteries," *J. Solid State Electrochem.*, vol. 16, no. 10, pp. 3307–3313, 2012.

[18] T. Ohzuku, A. Ueda, and N. Yamamoto, "Zero-Strain Insertion Material of Li[Li1/3Ti5/3]O4 for Rechargeable Lithium Cells," *J. Electrochem. Soc.*, vol. 142, no. 5, p. 1431, May 1995.

[19] B. Zhao, R. Ran, M. Liu, and Z. Shao, "A comprehensive review of Li4Ti5O12-based electrodes for lithium-ion batteries: The latest advancements and future perspectives," *Mater. Sci. Eng. R Reports*, vol. 98, pp. 1–71, 2015.

[20] C. P. Sandhya, B. John, and C. Gouri, "Lithium titanate as anode material for lithium-ion cells: a review," *Ionics (Kiel).*, vol. 20, no. 5, pp. 601–620, May 2014.

[21] N. Nitta, F. Wu, J. T. Lee, and G. Yushin, "Li-ion battery materials: present and future," *Mater. Today*, vol. 18, no. 5, pp. 252–264, 2015.

[22] T. Yuan, Z. Tan, C. Ma, J. Yang, Z.-F. Ma, and S. Zheng, "Challenges of Spinel Li4Ti5O12 for Lithium-Ion Battery Industrial Applications," *Adv. Energy Mater.*, p. 1601625, Jan. 2017.

[23] Y. Liu, S. Gorgutsa, C. Santato, and M. Skorobogatiy, "Flexible, Solid Electrolyte-Based Lithium Battery Composed of LiFePO4 Cathode and Li4Ti5O12 Anode for Applications in Smart Textiles," *J. Electrochem. Soc.*, 2012.

[24] X. Pu, L. Li, M. Liu, C. Jiang, C. Du, Z. Zhao, W. Hu, and Z. L. Wang, "Wearable Self-

Charging Power Textile Based on Flexible Yarn Supercapacitors and Fabric Nanogenerators," *Adv. Mater.*, vol. 28, no. 1, pp. 98–105, Jan. 2016.

[25] X. Wang, K. Jiang, and G. Shen, "Flexible fiber energy storage and integrated devices: recent progress and perspectives," *Mater. Today*, vol. 18, no. 5, pp. 265–272, Jun. 2015.

[26] J. G. Kim, D. Shi, M. S. Park, G. Jeong, Y. U. Heo, M. Seo, Y. J. Kim, J. H. Kim, and S. X. Dou, "Controlled Ag-driven superior rate-capability of Li4Ti5O12 anodes for lithium rechargeable batteries," *Nano Res.*, vol. 6, no. 5, pp. 365–372, 2013.

[27] H. L. Zou, H. F. Xiang, X. Liang, X. Y. Feng, S. Cheng, Y. Jin, and C. H. Chen, "Electrospun Li3.9Cr0.3Ti4.8O12 nanofibers as anode material for high-rate and low-temperature lithium-ion batteries," *J. Alloys Compd.*, vol. 701, pp. 99–106, 2017.

[28] B. Guo, Y. Li, Y. Yao, Z. Lin, L. Ji, G. Xu, Y. Liang, Q. Shi, and X. Zhang, "Electrospun Li4Ti5O12/C composites for lithium-ion batteries with high rate performance," *Solid State Ionics*, vol. 204–205, pp. 61–65, Dec. 2011.

[29] C. Liu, S. Wang, C. Zhang, H. Fu, X. Nan, and Y. Yang, "High power high safety battery with electrospun Li 3 V 2 (PO 4) 3 cathode and Li 4 Ti 5 O 12 anode with 95 % energy ef fi ciency," *Energy Storage Mater.*, vol. 5, pp. 93–102, 2016.

[30] H. Park, T. Song, H. Han, and U. Paik, "Electrospun Li4Ti5O12 nanofibers sheathed with conductive TiN/TiOxNy layer as an anode material for high power Li-ion batteries," *J. Power Sources*, vol. 244, pp. 726–730, 2013.

[31] L. Rayleigh, "On the Capillary Phenomena of Jets," *Proc. R. Soc. London*, vol. 29, no. 196–199, pp. 71–97, 1879.

[32] C. Wang, H. Koh, S. Ramakrishna, and S. Liao, *Electrospinning for Tissue Regeneration.* 2011.

[33] L. M. Duque Sánchez, L. Rodriguez, and M. López, "ELECTROSPINNING: LA ERA DE LAS NANOFIBRAS," *Rev. Iberoam. Polímeros Vol. Iber. Polímeros*, vol. 14, no. 141, pp. 10–27, 2014.

[34] J. V. Patil, S. S. Mali, A. S. Kamble, C. K. Hong, J. H. Kim, and P. S. Patil, "Electrospinning: A versatile technique for making of 1D growth of nanostructured nanofibers and its applications: an experimental approach," *Appl. Surf. Sci.*, 2017.

[35] Z.-M. Huang, Y.-Z. Zhang, M. Kotaki, and S. Ramakrishna, "A review on polymer nanofibers by electrospinning and their applications in nanocomposites," *Compos. Sci. Technol.*, vol. 63, no. 15, pp. 2223–2253, Nov. 2003.

[36] Y. Wu, M. V Reddy, B. V. R. Chowdari, and S. Ramakrishna, "Electrochemical studies on electrospun Li(Li1/3Ti5/3)O4 grains as an anode for Li-ion batteries," *Electrochim. Acta*, vol. 67, pp. 33–40, 2012.

[37] J.-F. Agassant, P. Avenas, P. J. Carreau, B. Vergnes, M. Vincent, J.-F. Agassant, P. Avenas, P. J. Carreau, B. Vergnes, and M. Vincent, "2 – Rheological Behavior of Molten Polymers," in *Polymer Processing*, 2017, pp. 33–175.

[38] A. K. Sharma, A. K. Tiwari, and A. R. Dixit, "Rheological behaviour of nanofluids: A review,"

Renew. Sustain. Energy Rev., vol. 53, pp. 779–791, Jan. 2016.

[39] N. Okutan, P. Terzi, and F. Altay, "Affecting parameters on electrospinning process and characterization of electrospun gelatin nanofibers," *Food Hydrocoll.*, vol. 39, pp. 19–26, 2014.

[40] M. Chowdhury and G. Stylios, "Effect of Experimental Parameters on the Morphology of Electrospun Nylon 6 fibres," *Int. J. Basic Appl. Sci. IJBAS-IJENS*, vol. 10, pp. 6–70.

[41] X.-Y. Wang, Y.-J. Li, C. Xu, L. Kong, and L. Li, "Synthesis and characterization of Li4Ti5O12 via a hydrolysis process from TiCl4 aqueous solution," *Rare Met.*, vol. 33, no. 4, pp. 459–465, 2014.

[42] M. Parhizkar, P. J. T. Reardon, J. C. Knowles, R. J. Browning, E. Stride, R. B. Pedley, T. Grego, and M. Edirisinghe, "Performance of novel high throughput multi electrospray systems for forming of polymeric micro/nanoparticles," *Mater. Des.*, vol. 126, pp. 73–84, 2017.

[43] D. Li and Y. Xia, "Electrospinning of Nanofibers: Reinventing the Wheel?," *Adv. Mater.*, vol. 16, no. 14, pp. 1151–1170, Jul. 2004.

[44] J. Pelipenko, J. Kristl, B. Janković, S. Baumgartner, and P. Kocbek, "The impact of relative humidity during electrospinning on the morphology and mechanical properties of nanofibers," *Int. J. Pharm.*, vol. 456, no. 1, pp. 125–134, 2013.

[45] T. Liang, M. Parhizkar, M. Edirisinghe, and S. Mahalingam, "Effect of humidity on the generation and control of the morphology of honeycomb-like polymeric structures by electrospinning," *Eur. Polym. J.*, vol. 61, pp. 72–82, 2014.

[46] B. Guo, Y. Li, Y. Yao, Z. Lin, L. Ji, G. Xu, Y. Liang, Q. Shi, and X. Zhang, "Electrospun Li4Ti5O12/C composites for lithium-ion batteries with high rate performance," *Solid State Ionics*, vol. 204–205, pp. 61–65, 2011.

[47] Y. Zhang, Y. Zhang, L. Huang, Z. Zhou, J. Wang, H. Liu, and H. Wu, "Hierarchical carambola-like Li4Ti5O12-TiO2 composites as advanced anode materials for lithium-ion batteries," *Electrochim. Acta*, vol. 195, pp. 124–133, 2016.

IOP Conf. Series: Materials Science and Engineering **437** (2018) 012017 doi:10.1088/1757-899X/437/1/012017

3D printing as a tool for predicting the mechanical-functional behavior of components and mechanical systems.

J M Argüello[1] and A Santos[1]

[1]Universidad Pontifícia Bolivariana, Bucaramanga, Colombia

juan.arguello@upb.edu.co

Abstract. Rapid prototyping is a technology that can automatically construct physical models from Computer Aided Design (CAD) data or is a group of techniques used to quickly fabricate a scale model of a physical part or assembly using three-dimensional computer aided design (CAD) data. This technic allows to minimize the possible human mistakes caused by a bad operator intervention, and represents a significant saving thanks to its accuracy, providing excellent visual aids for communicating ideas with co-workers or customers apart from design testing. The rapid prototyping area, where 3D models are converted into real models through different systems of manufacture, explores the possibilities offered by this technology for an engineer to design systems and devices before its final construction. As result it is observed an ostensible reduction in mechanical systems development times and its components; since functionality and usability of with rapid prototyping these corrections can be made. This document provides information about the use of Rapid Prototyping for specimens' preparation and general configurations for construction of functional models.

1. Introduction.

With the purpose of enhancing the teaching of the mechanical design, and thus facilitate the construction of knowledge and lead students to development of professional competences, were investigated modern technologies that are applied in the industry and that can be deployed as a new pedagogic strategy for the subjects of mechanical design.

Today the industry allows you to produce rapid prototypes to scale, in plastic PLA, from a Design of 3D modeling in CAD system.

The development of specific skills in students [1] leads to find an equivalence and association of concepts between the way of teaching the mechanical design and the design process that is done in the industry. In the processes of mechanical design, the production of prototypes is a practice that is becoming more and more entrenched. Why is spreading this activity? There are several reasons: accessible costs, existence of varied offer of software design by 3D modeling parametric, which based their logic in design through the geometry of the shapes and their parameters that can be varied according to need.

The student will be able to tour the design process through to analyze a specimen to understand the physical phenomena which act on it, enabling the development of reasoning physical - mathematician, and the understanding of these phenomena, or generate their own designs, simulate mechanically, to then be produced as a possible solution to a problem of real life.

The technology of Rapid Prototyping, thanks to its integration with the CAD and the development of New materials, opens infinite possibilities as a teaching resource in the teaching-learning process of mechanical engineering.

2. Process Methodology

This work will be developed in the following instances:

1) Study on the techniques of Rapid Prototyping, stages of the process and applications.
2) Manufacture of specimens to be analyzed
3) Conduct of stress tests to determine the behavior of each one of the configurations made
4) Realization of sampling by scanning electron microscopy.
5) Assessment of the behavior of each one of the configurations studied.
6) Proposal of pedagogic implementation for industrial purposes of a design exercise using the rapid prototyping.

2.1. Study on the techniques of Rapid Prototyping, stages of the process and applications

The process of rapid prototyping is the continuation of the modeling CAD. The product designer can quickly get a physical model of the design of a piece in place of a digital model or a drawing.

The prototype virtual, which is a modeling in computer of the design of the piece in a CAD software, may or may not be suitable for viewing the part. Using one of the technologies of Rapid Prototyping available, can be create a physical piece solid in a relatively short time, being able to examine visually and physically the workpiece, and then perform a series of tests and experiments to evaluate their advantages and disadvantages [2].

There are different technologies identified generically as rapid prototyping. They can be classified into two basic categories:

1) Processes of removal of material and

2) Processes of adding of material. The latter can be divided in turn according to the form of the starting material in the process of Rapid Prototyping:

a) Based on liquids,
b) Based on solid, and
c) Based on powders

The common approach in all current techniques of Rapid prototyping by adding material includes the following steps [3]:

1. Geometric modeling. Consists in modeling the piece in a system of CAD to define the overall volume. The solid modeling is the preferred technique because it provides a mathematical representation complete and accurate picture of the shape of the piece. For the rapid prototyping, the most important thing is to distinguish the interior (earth) on the part of its exterior and the solid modeling provides this distinction.

2. Tessellated of geometrical model. In this step, the model of CAD is converted to a format in which their surfaces are approximated by triangles or polygons. The triangles or polygons are used to define the surface, at least in an approximate way, and have their vertices arranged in such a way that you can distinguish between the interior of the object of its exterior. The format of Tessellated common that is used in the rapid prototyping is STL (Standard Triangle Language), which has become the de facto standard as input format for almost all the systems of rapid prototyping.

3. Division of the model on layer. In this step, the model in the form of STL file is divided into horizontal layers parallel with a separation very close. These layers are used by the system of Rapid Prototyping to build the physical model. By convention, the layers are formed in the orientation of the x-y level, and the procedure of creating layers occurs in the direction of the z axis as can be seen in Figure 1.

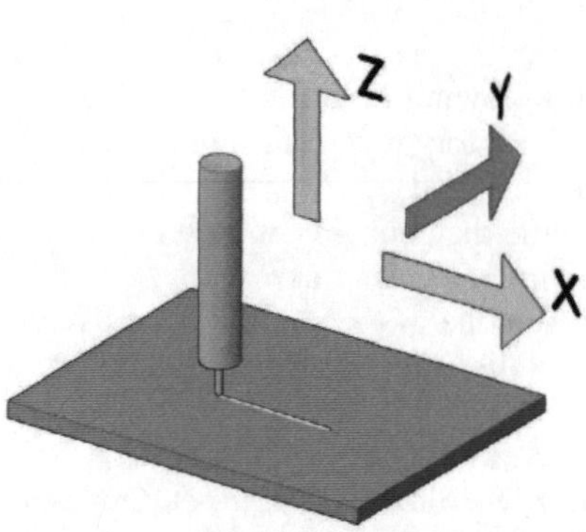

Figure 1. Scheme of the modelled by Fused Deposition Modelling.
http://web.fceia.unr.edu.ar/images/PDF/secretaria_desarrollo_institucional/segunda_jornada_exp_innov_en_educacion_fceia/Bortolato-Lenti-Morelli-Verger.pdf

In this work, we will address the modeling by Fused Deposition Modeling (FDM).

It is a process of Rapid Prototyping in which a filament of polymer is stretched over the surface of the existing part from a head of work to complete each new layer. The work head is controlled in the x-y level during each layer and then moves up a distance equal to one layer in the z direction. The starting material is a filament robust fed from a spool to the work head that heats the material to a temperature 0.5° C above their melting point before stretching on the workpiece surface (Figure 2). The material stretched solidifies and welded to the surface of the part that this colder in about one-tenth of a second. The form of work is similar to a plotter 3D (Figure 3). If you need support, uses a media separable easily from the material of modeling parent.

Figure 2. work head
https://www.makerbot.com/img/replicator/smart_extruder_module_7.jpg

Figure 3. MAKERBOT REPLICATOR+
https://www.makerbot.com/replicator/

In the case of study, both the starting material as the support will be a thermoplastic polymer called acid polyacrylic (PLA), which is obtained from maize starch (U.S.) or cassava (mostly in Asia), or sugar cane (rest of the world), taking properties similar to those of polyethylene terephthalate (PET) and is used for the manufacture of packaging, but that in addition is biodegradable, deteriorate easily in water and carbon monoxide.

2.2 Manufacture of specimens to be analyzed
For the study of characterization, have been printed15 specimens in a printer Makertbot of modeling by FDM , divided into three groups (low, medium and high) as print resolution. In Figure 4 you can see the weight in grams of each one of them; where each layer has a layer height of 0.3 / 0.2 / 0.1 mm, respectively; where a lower height will be better print quality.

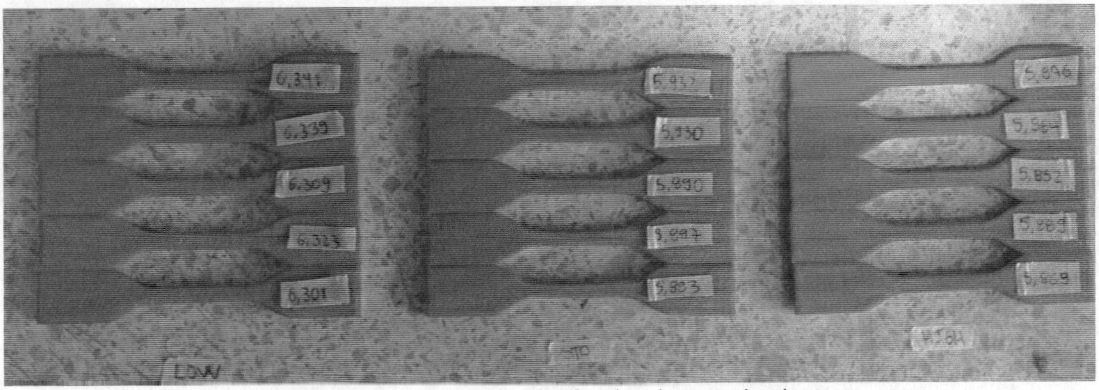

Figure 4. Test specimens for the characterization

The print resolution is directly proportional to the print time; at a higher resolution, the longer the time that lasts the printing; this must also be taken into account at the time to generate the print models and the function to fulfill. For the case study, each test piece lasted printed between 30 and 45 minutes.

2.3 Conduct of stress tests to determine the behavior of each one of the configurations made
15 printed specimens were subjected to a stress test to take them to their breaking point; starting with the of lower resolution (Figure 5), which reaches a point of break with 0.044 KN/mm2; to the of higher resolution (Figure 6), which reaches its point of break with 0,047 KN/mm2.

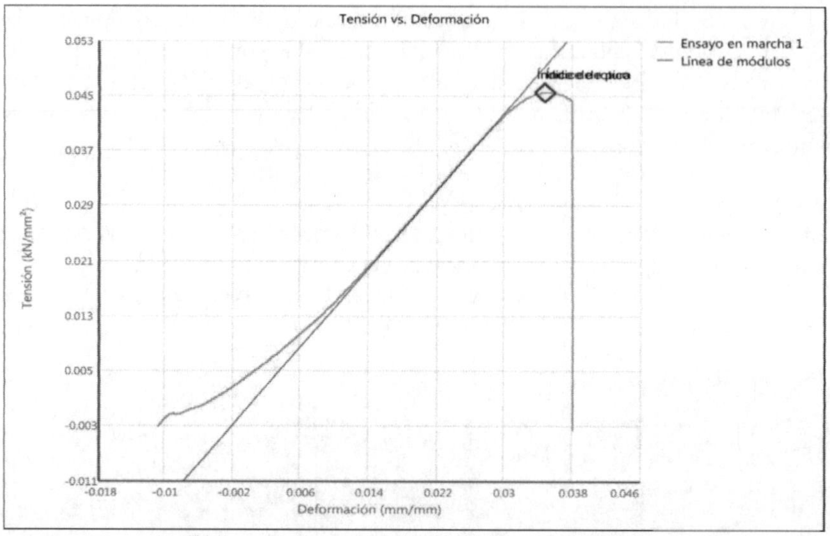

Figure 5. Stress test to the test specimen 1

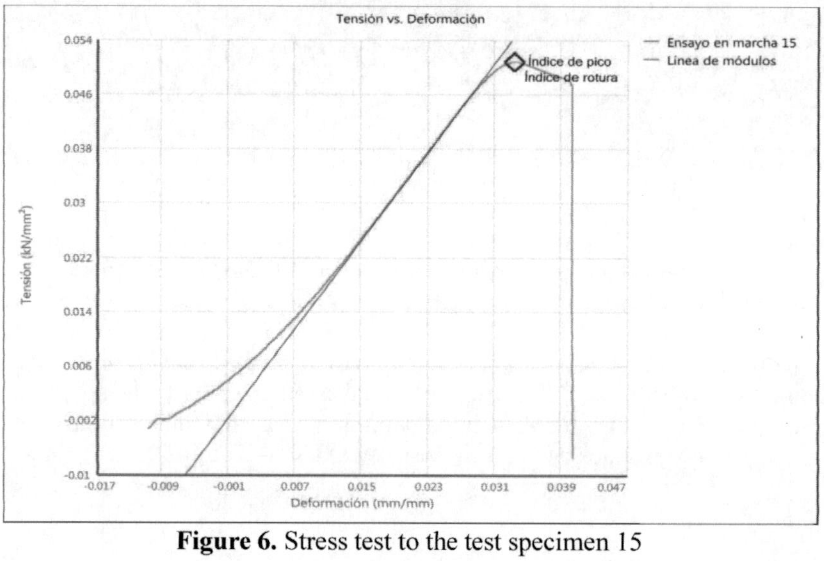

Figure 6. Stress test to the test specimen 15

Later, to every specimen a study was done to him in the electronic microscope of sweep SEM. The electronic microscopy of sweep or SEM is based on the beginning of the optical microscopy in which the light bundle is replaced with an electrons bundle [4]. In the above-mentioned study, it was possible to characterize of microscopic form the morphology of each of them, to understand how to major impression resolution, major layers quantity is observed, what it does to the most resistant specimen in the tension essay.

2.4 Realization of sampling by Scanning electron microscopy
In the figure 8 one can observe the disposition of the layers of the test tubes of half resolution. In the figure 7 one can observe the disposition of the layers of the test tubes of half resolution.

IOP Conf. Series: Materials Science and Engineering **437** (2018) 012017 doi:10.1088/1757-899X/437/1/012017

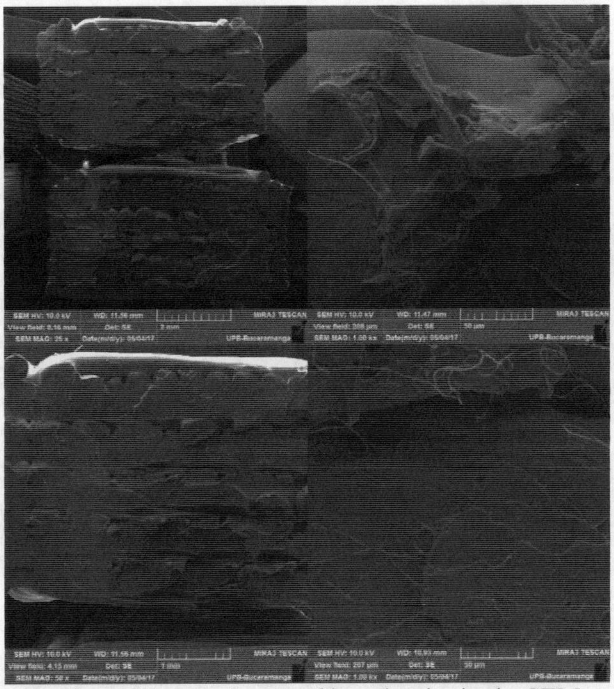

Figure 7. Characterization of the specimens of low density in the SEM scanning electronic microscope

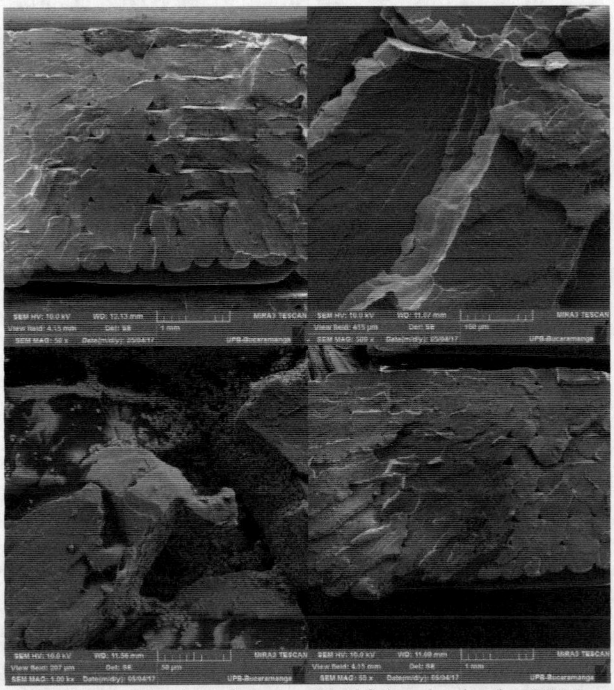

Figure 8. Characterization of the specimens of half density in the SEM scanning electronic microscope

IOP Conf. Series: Materials Science and Engineering **437** (2018) 012017 doi:10.1088/1757-899X/437/1/012017

In the figure 9 one can observe the disposition of the layers of the test tubes of high resolution.

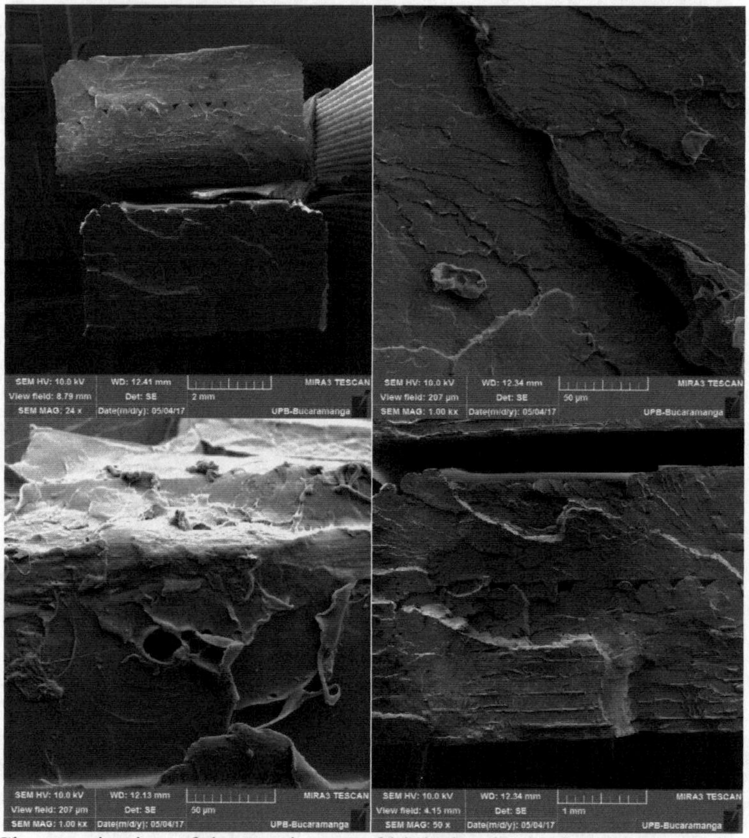

Figure 9. Characterization of the specimens of high density in the SEM scanning electronic microscope

In photographs taken by the microscope SEM of printing at low resolution you can appreciate the deposition of 6 layers; in the medium-resolution there are also 9 layers; while in the high resolution presented 12 layers. Which is the reason for which makes it more resistant the measuring cylinder to increase the print resolution.

2.5 Assessment of the behavior of each one of the configurations studied

Making a comparison between the different types of print resolution, with regard to the effort applied that resist before fail and the characterization carried out in the test specimens in the microscope SEM; it can be inferred that can be modeled and print prototypes, preliminary or final; functional or formal; that could be useful in the study of a piece and / or product, applying the concepts of mechanical design, with the possibility to study it and submit it to different tests according to interest concerned. This means that when it is necessary to perform a conceptualization morphology of the piece, that practice would be to make a print at low resolution (Figure 10); when you perform a study of functional type in terms of movement and assembly, ideally make a print with a medium resolution (Figure 11); finally, if the need is to determine the behavior of a piece and / or assembly with some mechanical loads, and interaction repetitive, it is therefore more appropriate to perform printing in high resolution (Figure 12).

Figure 10. Hydraulic jack printed at low resolution to appreciate the relationship between parts and present it as a formal model

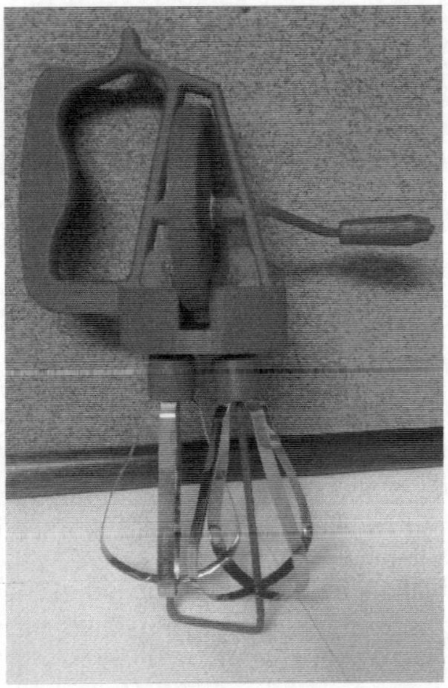

Figure 11. Manual mixer of food that will be subjected to interaction with the user and other materials, printed with medium resolution

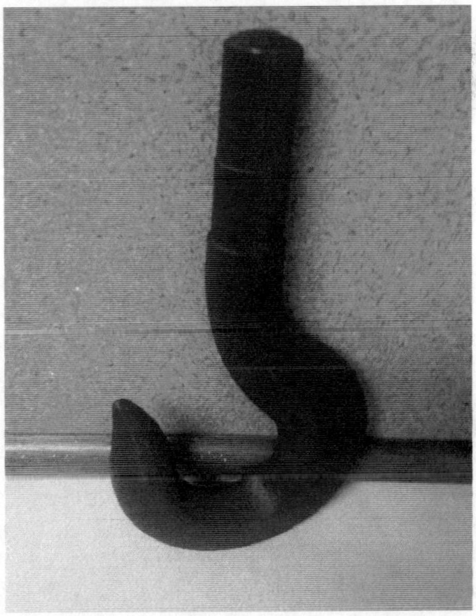

Figure 12. Hook crane that will be subjected to small mechanical loads, as demonstration to the students of design applied 2, printed in high resolution

It is important to emphasize that the printing speed and the consumption of material is directly proportional to the print resolution; therefore, for a same model printed in a low resolution, the quantity of material consumed could be 70%, and the time spent would be a 60% if compared with the same model printed in high resolution.

2.6 Proposal of pedagogic implementation for industrial purposes of a design exercise using the rapid prototyping

The new paradigms educational, as the focus of the student-centered learning and training based on competencies, which at present is an international trend in the design of plans of engineering study [1], requires of the teacher the implementation of new pedagogic strategies and resources directed to the achievement of significant learning in students, forming the future engineers that society demands today.

In regard to the Mechanical Engineering, the introduction of CAD systems, from some 25 years ago, meant a radical transformation of the "way of thinking and doing" in all areas relating to mechanical design and graphic communication. The graphics technology of CAD systems not only impacted on the design of curricula but that imposed a new geometry digital. The possibility of 3D solid modeling on the basis of the 3D geometry of the objects and, from that modeling, obtaining the views automatic in the plane or projections, reversed the traditional sequence of the design process, from 3D to 2D, which meant a major paradigm shift in the area of graphic representation [5]; it made possible a direct interaction, fast and economic of the students with the mechanical elements designed (Figures 10, 11 and 12).

The production of prototypes in the process of mechanical design is a practice that tends to become widespread in the industry. Technological tools such as CAD-CAM (design and manufacturing assisted by computer), or the CNC (computerized numerical control) produce models or prototypes directly from the interface design in a computing environment CAD 3D; therefore, the display multi-sensory (visual-Touch) constitutes a source of information and understanding of the reality that provides great help in the process of mechanical design and representation and graphic interpretation. In the development of pieces or complex objects can be much more information the model put it in the hand that watching simply on the monitor of a computer. The student will be able to analyze the geometry of its form, check its physical behavior and mechanical, experience movements and possibilities of assemblies. It is here where research and teaching must be found, since it is necessary to investigate to teach and to teach investigating.

3. Conclusions

The incorporation of technology of Rapid Prototyping, comes from the productive process in the industry, in the teaching-learning process in the area of mechanical design, enables the transfer of experiences of the industrial activity in the classroom, and therefore constitutes an educational innovation that is also in line with the current concept of training based on competencies.

As well as the solid modeling 3D facilitates the design of complex shapes and their graphical representation, Rapid Prototyping enables easy implementation of a formal model or a functional model, and power ability deductive and spatial reasoning of the student. The cost of production of models by Rapid Prototyping is no longer a limiting factor for its use extended in fields such as education, because today this technology is much more accessible than in its beginnings, some 20 years ago.

References

[1] MORELLI, R., Lenti, C. (2013). Educación basada en competencias para el área de Sistemas de Representación. Libro Resúmenes VI CAEDI: "Congreso Argentino de Enseñanza de la Ingeniería", Salta, Sep. 2013, ISBN 978-987-633-012-1, 12.

[2] MCLEOD Peter (2001). Technology Watch. PRIME Faraday Partnership. Wolfson School of Mechanical and Manufacturing Engineering. Loughborough University, Loughborough, Leics LE11 3TU. ISBN 1-84402-018-5

[3] GROOVER, M. P. (2013). Fundamentos de manufactura moderna. 3^{Ed}. Ed. McGraw Hill, México, ISBN 978-970-10-6240-1.

[4] La microscopía electrónica de barrido SEM (I) Concepto y usos. https://www.patologiasconstruccion.net/2012/12/la-microscopia-electronica-de-barrido-sem-i-concepto-y-usos/

[5] FLORIO, W., LASAR Segall, M., SOARES DE ARAÚJO, N. (2014) A contribuição dos protótipos rápidos no processo de projeto em arquitetura. Caderno de Resumos: VII International Conference on Graphics Engineering for Arts and Design - XVIII Simpósio Nacional de Geometria Descritiva e Desenho Técnico. GRAPHICA 2014, Curitiba, Paraná, Brasil, nov. 2014, ISBN 978-85-61172-00-8, 159.

AUTHOR INDEX

Institute of Physics
Dirac House, Temple Back
Bristol BS1 6BE UK

ISSN: 1757-8981
ISBN 978-1-5108-7655-2

CURRAN ASSOCIATES INC.
proceedings.com